CATASTROPHES
in EARTH HISTORY

A Source Book of
Geologic Evidence,
Speculation and Theory

by

Steven A. Austin, Ph.D.

ICR Technical Monograph 13

Institute for Creation Research
El Cajon, California

CATASTROPHES IN EARTH HISTORY: A Source Book of Geologic Evidence Speculation and Theory -- Technical Monograph 13

Copyright © 1984

Institute for Creation Research
2100 Greenfield Drive
El Cajon, California 92021

ISBN No. 0-932766-08-0

Library of Congress Catalog Card Number 83-080181

ALL RIGHTS RESERVED

No portion of this book may be used in any form without written permission of the publisher, with the exception of brief excerpts in magazine articles, reviews, etc.

Cataloging and Publication Data

Austin, Steven A.
  Catastrophes in earth history: a source book of geologic evidence speculation and theory
(Technical Monograph, No. 13)

  1. Catastrophes (Geology)
2. Earth--Processes. 3. Geology--Physical Processes. I. Title
        551.7
Printed in the United States of America

PREFACE

One thing is sure to incite emotional response from geologists, geophysicists and geomorphologists--it is the idea of catastrophes in earth history. Just propose that a regional or global catastrophe left evidence in the geologic record and you will be promptly charged with giving free reign to fantasy. Even worse, your notions may be assigned to a shelf with a whole gamut of suggestions ventured by innumerable "crackpots." Such neglect or censorship has existed in geological science for the last 150 years when "megathinking" was renounced, and "microthinking" was the standard. However, the situation has changed. Geology contains a rich body of evidence, speculation and theory challenging the notion that the earth evolved to its present configuration simply by the action of gradual processes. This book is about some extraordinary processes that have been discovered, and which challenge our way of thinking about the earth.

I began collecting the material for this volume sixteen years ago. Later, the topic became my master's thesis in geology. In the work of reviewing thousands of volumes of geologic literature to collect the material for this source book, I have been graciously assisted by students, professional geologists, librarians and secretaries. The manuscript was composed, revised, and indexed on a word processor. Many publishers kindly granted me permission to reprint sections from copyrighted publications. I originally intended this book to be a resource for graduate students to help them ask questions they would never have thought to ask. I am also gratified to know that my collection of abstracts and quotations is of general interest to scientists and laymen alike.

ACKNOWLEDGEMENTS

Permission to quote significant portions of copyrighted material has been obtained from many sources including:

American Association for the Advancement of Science
American Association of Petroleum Geologists
American Journal of Science
California Division of Mines and Geology
Cambridge University Press
Elsevier Scientific Publishing Company
Geoscience Research Institute
John Wiley & Sons
KRONOS, A Journal of Interdisciplinary Synthesis
McGraw-Hill Book Company
Mountain Press
National Geographic Society
New Science Publications
Pergamon Press
Royal Netherlands Academy of Sciences
Scottish Academic Press
Sigma Xi, the Scientific Research Society
Society of Economic Paleontologists and Mineralogists
Texas Bureau of Economic Geology
University of Texas at Austin, Bureau of Economic Geology
University of Texas Press
W. H. Freeman and Company
William Morrow and Company

Many of the unusual photographs in this book were kindly provided by the National Atmospheric and Space Administration (NASA) and by the National Geophysical Data Center of the National Oceanic and Atmospheric Administration (NOAA).

TABLE OF CONTENTS

Preface

Acknowledgements

Chapter 1 - About the Book.................................... 1

Chapter 2 - History and Philosophy............................ 7

    The Problem Stated.. 9
    Definitions and History................................... 16
    The Failings of Uniformitarianism......................... 27
    The Opposition to Catastrophism........................... 31

Chapter 3 - Cosmic Catastrophes............................... 37

    The Tunguska Explosion of 1908............................ 38
    Probability of Asteroid Impact............................ 43
    Physics of Impact with Rock............................... 48
    Ancient Impact Evidences.................................. 51
    Tektites.. 58
    Effects of Impact... 62
    Effects of Supernovae..................................... 68

Chapter 4 - Extrusive and Intrusive Catastrophes.............. 71

    Volcanoes in History...................................... 72
    Volcanic Structures....................................... 91
    Prehistoric Pyroclastic Deposits.......................... 96
    Prehistoric Lava Flows.................................... 106
    Hot Intrusive Processes................................... 114
    Cold Intrusive Processes.................................. 122
    Sedimentary Products of Volcanism......................... 127

Chapter 5 - Mass Movement Catastrophes........................ 135

    Historic Gravity Flow Deposits............................ 136
    Prehistoric Gravity Flow Deposits......................... 140
    Historic Gravity Slide Deposits........................... 148
    Prehistoric Gravity Slide Deposits........................ 152
    Historic Earthquakes and their Effects.................... 160
    Prehistoric Earthquakes and their Effects................. 167
    Collapse Features... 177

Chapter 6 - Water Catastrophes................................ 173

    Historic Water Catastrophes............................... 174
    Water Catastrophes and the Origin of Sedimentary Rocks.... 198
    Catastrophic Burial in Fossilization...................... 208
    Ancient Geomorphic Features as Evidence of Water
        Catastrophes.. 216
    Physics of Asteroid Impact with Water..................... 226

Chapter 7 - Atmospheric Catastrophes.......................... 231

    Air Waves... 232
    Eruption and Explosion of Gases........................... 236
    Climatic Effects of Catastrophic Volcanism................ 241
    Glaciation.. 245
    Changes in Atmospheric Composition........................ 252

Chapter 8 - Related Topics.................................. 257

   Rapid Lithification and Fossilization..................... 258
   Extinction and the Fossil Record.......................... 266
   Features Attributed to Slow Formation
      Which May Indicate Rapid Formation.................... 271
   Catalogs of Catastrophic Processes........................ 285

Author Index.. 291

Subject Index... 301

Chapter 1

ABOUT THE BOOK

Every ten minutes an earthquake can be felt somewhere in the world, and one in fifty will inflict damage. Fifteen earthquakes each year may unleash the explosive force of a million tons of TNT, but only one may cause major change in the form of the land surface. History records many devastating earthquakes, but could they occur with enough energy to dislodge a mountain range from the earth's surface? Might catastrophic earthquakes of violence unknown in human experience have reshaped and helped to form the earth we see?

A hundred lightning bolts strike the earth every second and countless billions of raindrops fall from the sky each moment. Although history records a mysterious extraterrestrial, atmospheric explosion over Siberia in 1908 possessing the energy of a large hydrogen bomb, humans have not witnessed a single crater excavated by the fall of an asteroid or comet. Has the earth been struck by objects from outer space with enough force to cause the extinction of species? Could an extraterrestrial object's collision with the ocean generate a sea wave which might inundate a continent?

This leads to a more fundamental question: **how does the earth change**? Does the earth always change under the action of processes which operate at essentially the same rate, scale or intensity that have been observed today--as uniformitarian geologists maintain? Or, does the earth change on occasion by the operation of natural processes rapidly and catastrophically, interposed between periods of slow and gradual change--as catastrophist geologists suggest? Are the insignificant **and** the extraordinary the architects of the modern world?

Modern geology contains a rich body of research and speculation that directly challenges the theory that the earth evolved to its present configuration only by gradual processes. This book is a compendium of geologic literature documenting catastrophes and catastrophic processes some of which have affected the earth on a regional and global scale. The literature from a variety of scientific fields is cited, some by direct quotation, and some by abstracting the

---

A **catastrophe** may be defined as a natural event of large magnitude (energy), short duration, wide extent and low frequency. The word <u>catastrophe</u> comes from the Greek <u>kata</u>, meaning "thoroughly," and <u>strophe</u> meaning "turning."

About the Book

essence of what the author said. In many of the abstracts "plain English" is used to state what the author communicated in a more technical fashion. Included in many of the reviews of the literature are definitions of terms not supplied by the author and discussion of the relationship of the author's statements to the literature and themes of this book. A balance has been sought between modern and ancient catastrophic processes. The usefulness of the book is enhanced by topical organization and an exhaustive index of both authors and subjects.

The 249 publications cited represent the work of 346 authors. In selecting each of the references on catastrophic geologic processes for this book four questions were asked.

1. Does the author report original observations or in other ways demonstrate first hand knowledge of the subject being discussed?

2. Is there evidence that the author has training and qualification for analysis of the material presented?

3. Have the author's statements influenced the general body of knowledge that they address?

4. Have the author's statements received scholarly review by

competent authority prior to publication?

If at least three of the above questions could be answered in the affirmative, the publication was judged to be worthy to be abstracted. No doubt, hundreds of other works not cited in this book might be included also. Although only sources available in the English language are cited here, a wide variety of translated articles are included. Catastrophist geology is not confined to English-writing geologists.

Caution needs to be exercised in using this book. First, no claim is made as to the accuracy of the catastrophic events that are inferred from the geologic data. The interpretations of scientists are simply reported so that the reader can be aware of the range of subjects under discussion, and so the reader can realize that in specific subject areas difficulties have been encountered which appear to require catastrophic explanations. Second, the reader may need to become familiar with the geologic subjects being discussed so that the statements can be interpreted correctly in the context in which they were made. The reader is encouraged to obtain adequate background knowledge, then to consult the original references cited in this book. Third, although there is much here to

About the Book

stimulate the imagination, the reader is cautioned to refrain from speculation, except where tests or further research can be devised. Letting the evidence speak and developing a methodology of investigation are essential to good science.

To what can we liken catastrophist geology? It is like a grand, stately forest which long ago experienced a devastating fire, and which, now, is beginning to regrow with unrestrained vigor. That fire, of course, was the stagnant quagmire of uniformitarian orthodoxy which, in the late nineteenth and early twentieth centuries, appeared to have killed everything exceptional and unique. In this regard we might also mention another fire--the fire of imagination--and the admonition of Plutarch, "The mind is not a vessel to be filled but a fire to be lighted." Few topics stimulate the imagination more than catastrophic events in earth history. This book is a guide to serve in the analysis of some of the earth's most challenging mysteries. Study of the extremes in Nature should not just astonish us; it should challenge us to understand the full range of processes which have formed the earth. The sage words of Sherlock Holmes apply to geology:

> ". . . when you have eliminated the impossible, whatever remains, however improbable, must be the truth."

Chapter 2

HISTORY AND PHILOSOPHY

Historians and philosophers of science have argued that science is built upon paradigms which have broad acceptance. To change from one paradigm to another is quite difficult because there is much inertia to be overcome. Individually, we cherish concepts also. The earth is our home. To suggest that the earth has been deluged by water, impacted by asteroids, and devastated by volcanoes brings up a host of scientific, philosophical, ethical and religious issues. We naturally possess biases in these areas and may even react emotionally when confronted with such ideas. Paleontologists, for example, have been most reluctant to entertain notions of geologic catastrophes to explain biologic extinctions. Professor David M. Raup, Chairman of Geophysical Sciences at the University of Chicago, said, "Perhaps the most emotional aspect of all this is that catastrophism has to be included in our models of evolution. Paleontologists and evolutionary biologists have been absolutely locked into a dogma of gradualism (slow, steady and constant change) . . ." (Los Angeles Times, Sep. 4, 1983). History demonstrates that the success of investigations of ancient catastrophes hinges upon presuppositions and methods of investigation. This chapter begins by stating the problem of interpreting ancient catastrophes (references 1 through 5), defines terms in their historical context (references 6 through 14), discusses the failure of uniformitarianism (references 15 through 19), and exposes the popular bias of geologists toward catastrophism (references 20 through 27).

The Problem Stated

1 Kloosterman, J. B., 1976, Catastrophist geology: Catastrophist Geology, vol. 1, no. 1, pp. 1-3.

Uniformitarianism holds that the processes governing the Earth's organic and inorganic past were the same as those apparent today, and that they operated then at the same intensity and rate as now. When they consider this definition thoughtfully, many geologists realize that they do not really agree with it. Too many events in the Earth's history do not fit a uniformitarian system--enormous calderas, plateau basalts, ice ages, alpine nappes, bone breccias, the sudden appearance of diversified life at the close of the Precambrian, the abrupt extinction of dinosaurs and ammonites, and so on. In a uniformitarian system the sedimentological and paleontological records are contradictory; if we assume uninterrupted sedimentation, we have to accept catastrophes in evolution; if we do not accept catastrophes in evolution we have to postulate major gaps in the sedimentary record.

Catastrophism admits the occurrence of discontinuities in Earth history--because we observe them now and because we are forced to infer them from the geological record. Even such Lyellian agents as the raindrop and the sand grain often do their work in discontinuous manners: the catastrophic erosion after a lakespill for example, or sedimentation by turbidity current. In spite of our proclaimed uniformitarianism, catastrophist hypotheses abound-- the capture of the Moon, astroblemes, bursts of cosmic rays, natural nuclear reactors, the breaking up and the collision of continents. When proposed by geologists of non-catastrophist persuasion, such hypotheses are taken seriously, but when similar ideas are forwarded by less conditioned outsiders, they are regarded as evidence of lunacy simply because they violate uniformitarian dogma. Mainstream geologists often do not even try to formulate clearly their own ideas; while they are cheating, somebody relegated to the lunatic fringe may be exposing the fraud.

Catastrophes do occur. The dinosaurs did die out--whether it took a million years or a day-- either through the cumulative effect of continuous causes, actualistic or not, or through a unique, sudden, terrestrial or extraterrestrial event. Should such riddles ever be solved, the solutions will come from an inspired search for clues and not through application of the methods of medieval scholastics or nineteenth-century rationalists. (pages 1 and 2)

2 Heylmun, E. B., 1971, Should we teach uniformitarianism?: Journal of Geological Education, vol. 19, pp. 35-37.

Many geology and earth-science instructors accept and teach the doctrine of uniformitarianism with little further thought since it is considered to be one of those basic "laws" that form the very foundation of the geological sciences ... Few geologists are willing to accept ideas of major, world-wide cataclysms, as the catastrophists were supposedly dismissed over two-hundred years ago. Many textbooks present the doctrine in such a way as to leave the reader with the impression that it is an indisputable "law." ...

It is hereby submitted that most scientists are guilty of an overly-zealous interpretation of the doctrine of uniformitarianism. Many instructors dismiss the possibilities of global catastrophes altogether, whereas others ridicule and scoff at the early ideas. These same instructors will implore their students to think scientifically and to develop the principles of multiple-working hypotheses. The fact is, the doctrine of uniformitarianism is no more "proved" than some of the early ideas of worldwide cataclysms have been disproved. (page 35)

3 Ager, D. V., 1981, The nature of the stratigraphical record: New York, John Wiley, second edition, 122 pp.

An experienced geologist delivers a compelling critique of modern uniformitarian geology. Speaking of geologists, Ager says,

. . .we have allowed ourselves to be brainwashed into avoiding any interpretation of the past that involves extreme and what might be termed 'catastrophic' processes. (page 46)

He says that the science went astray when ". . .geology got into the hands of the theoreticians who were conditioned by the social and political history of their day. . ." (page 46). The theoreticians, he says, were the early uniformitarians, whereas the catastrophists were careful field observers. Ager presents seven catastrophist principles of geology:

1. The "phenomenon of the persistence of facies" (at certain times particular sedimentary environments were prevalent over vast areas of the earth),

2. The "phenomenon of the fallibility of the fossil record" (paleontologists err by mesuring the past in terms of the present)

3. The "phenomenon of the gap being more important than the record" (the sedimentary rock record is a fragmentary record of the earth's history),

4. The "phenomenon of the catastrophic nature of much of the stratigraphical record" (sedimentation in the past has been very rapid and spasmodic),

5. The "phenomenon of quantum sedimentation" (periodic catastrophic events may have more effect than vast periods of gradual evolution),

6. The "principle of the relative independence of sedimentation and subsidence" (sedimentation does not usually occur at the same rate as subsidence),

7. The "principle of the golden spike" (stratigraphic unit must be arbitrarily decided upon, then accepted, so that geologists can proceed to more important matters).

Ager concludes his book using a metaphor comparing the history of the earth to the life of a soldier. Both, he says, experience "long periods of boredom and short periods of terror."

4 Russell, D. A., 1982, The mass extinctions of the late Mesozoic: Scientific American, vol. 246, no. 1, pp. 58-64.

> Catastrophism is not a new doctrine in efforts to account for episodes in the history of the earth, but it has not been a particularly popular one. Early in the 19th century, when geology was in its infancy, the French anatomist Georges Cuvier suggested that the past had been marked by a series of environmental "revolutions," or catastrophes. In his view such disruptions would account for three animal disappearances: that of the mammoths at the end of the ice age, that of the many primitive mammals fossilized in rocks lying deeper than the ice-age gravels and that of the giant reptiles fossilized in chalk beds lying deeper still. In the decades that followed, however, the work of such pioneer geologists as Charles Lyell made it apparent that the processes of change in earth history were of far greater duration than Cuvier had believed.

Catastrophism fell from favor, to be replaced by the doctrine of gradualism. For more than a century now paleontologists have generally agreed that whatever may have caused the disappearances at the end of the Mesozoic era, it could not have been a worldwide catastrophe. (page 58, Copyright 1982 by Scientific American, Inc. All rights reserved.)

Russell, a paleontologist, then reports evidence for the extinction of dinosaurs by a global catastrophe. He admits the bias which has existed toward catastrophism and vindicates Cuvier.

5 Dury, G. H., 1980, Neocatastrophism? A further look: Progress in Physical Geography, vol. 4, pp. 391-413.

This paper takes further an inquiry (Dury, 1975) into the prospects of a powerful growth of neocatastrophism, whether under the actual name or not. Coined for the purposes of paleontology, the term has already escaped into geoscience in general. The concept of the significance, and in some connections the dominance, of events of great magnitude and low frequency is gaining ground among sedimentologists at least. But the words neocatastrophism and neocatastrophist carry strong emotional overtones, as has been revealed in the renewed debate about uniformitarianism. (page 391)

Definitions and History

History and Philosophy

6 Bates, R. L., and Jackson, J. A., eds., 1980, Glossary of geology: Falls Church, American Geological Institute, second edition, 749 pp.

The authoritative source of definitions in geology defines a "catastrophe" as "a sudden, violent disturbance of nature, ascribed to exceptional or supernatural causes, affecting the physical conditions and the inhabitants of the Earth's surface" (page 98). None of the literature in this bibliography refers to "supernatural causes" and it appears unnecessary to include it in the definition. A better definition is that a catastrophe is a natural event of large magnitude (energy), short duration, wide extent and low frequency.

7 Austin, S. A., 1979, Uniformitarianism--a doctrine that needs rethinking: Compass, vol. 56, pp. 29-45

This paper attempts to define the term uniformitarianism, trace its history, and assess its status in modern geology. Charles Lyell's doctrine contains four concepts: (1) a methodological principle asserting temporal continuity of the properties of matter and energy as described by scientific laws, (2) a causal principle requiring temporal continuity of the kinds of geological processes, (3) an actional theory affirming temporal uniformity of rates of geological processes, and (4) a configurational theory alleging temporal uniformity of geological conditions. During the last one hundred fifty years, uniformitarianism has been defined in various ways by different geologists depending on which of Lyell's four concepts is accepted as true--hence the semantic confusion of as many as four concepts under the single term. Although the affirmation of the continuity of the properties of matter and energy in Lyell's methodological principle is good scientific method, inclusion of it under the term uniformitarianism is superfluous because it is already asserted when it is said, "Geology is an inductive science." Lyell's causal principle is false if understood to require temporal continuity of all modern geological processes, and only known modern processes; it is dissolved by a principle of simplicity if understood as a procedural statement, not limiting ancient processes, but only the geologist's accounting of those processes. Lyell's two theories affirming temporal uniformity of rates of processes and uniformity of geological conditions have been refuted by geological data, which, alone, must be used to understand ancient rates and conditions. The term uniformitarianism, therefore, should be abandoned when describing formal assumptions used in modern geological inquiry. (page 29)

8 Lyell, C., 1875, Principles of geology; or, the modern changes of the earth and its inhabitants considered as illustrative of geology: London, Murray, twelfth edition, two vols., 655 pp. and 652 pp.

Charles Lyell, the popularizer of uniformitarian geology, considered catastrophism a "dogma" and rejected ideas of global catastrophes.

> Never was there a dogma more calculated to foster indolence, and to blunt the edge of curiosity, than this assumption of the discordance between the ancient and existing causes of change. It produced a state of mind unfavourable in the highest degree to the candid reception of the evidence of those minute but incessant alterations which every part of the earth's surface is undergoing.
>
> For this reason all theories are rejected which involve the assumption of sudden and violent catastrophes and revolutions of the whole earth, and its inhabitants--theories which are restrained by no reference to existing analogies, and in which a desire is manifested to cut, rather than patiently to untie, the Gordian knot. (Vol. I, page 318)

9 Lyell, K. M., 1881, Life, letters and journals of Sir Charles Lyell: London, John Murray, 2 vol., 475 pp. and 489 pp.

In a letter to Roderick Murchison written in 1829, Charles Lyell states the purpose of his soon-to-be-published book <u>Principles of Geology</u>:

> My work is in part written, and all planned. It will not pretend to give even an abstract of all that is known in geology, but it will endeavour

> to establish the principle of reasoning in the science; and all my geology will come in as illustration of my views of those principles, and as evidence strengthening the system necessarily arising out of the admission of such principles, which, as you know, are neither more nor less than that no causes whatever have from the earliest time to which we can look back, to the present, ever acted, but those now acting; and that they never acted with different degrees of energy from that which they now exert. (vol. I, page 234)

10 Whewell, W., 1832, [Review of Principles of Geology]: Quarterly Review, vol. 47, no. 93, pp. 103-132.

William Whewell, the polymath of the early nineteenth century catastrophist school, wrote an extensive review of the first edition of Charles Lyell's book Principles of Geology. This review contains the first use in print of the words "Uniformitarians" and "Catastrophists." Concerning Lyell's book, Whewell says:

> With this striking exception, we may assert, with our author and other geologists, that all the facts of geological observation are of the same kind as those which occur in the common history of the world. The question then comes before us--are the extent and the circumstances of the geological phenomena of the same order as those of which the evidence has thus been collected? Have the changes which

History and Philosophy

> lead us from one geological state to another been, on a long average, uniform in their intensity, or have they consisted of epochs of paroxysmal and catastrophic action, interposed between periods of comparative tranquillity?
>
> These two opinions will probably for some time divide the geological world into two sects, which may perhaps be designated as the <u>Uniformitarians</u> and the <u>Catastrophists</u>. The latter has undoubtedly been of late the prevalent doctrine, and we conceive that Mr. Lyell will find it a harder task than he appears to contemplate to overturn this established belief. Indeed, we think it ought to be so. It seems to us somewhat rash to suppose, as the uniformitarian does, that the information which we at present possess concerning the course of physical occurrences, affecting the earth and its inhabitants, is sufficient to enable us to construct classifications, which shall include all that is past under the categories of the present. Limited as our knowledge is in time, in space, in kind, it would be very wonderful if it should have suggested to us all the laws and causes by which the natural history of the globe, viewed on the largest scale, is influenced--it would be strange, if it should not even have left us ignorant of some of the most important of the agents which, since the beginning of time, have been in action; of something, in short, which may manifest itself in great and distant catastrophes. (page 126)

Whewell's comments appear prophetic. The debate between uniformitarians and catastrophists has persisted now for 150 years, surviving because of the reasons he mentions!

11 Conybeare, W. D., 1830, [Letter to editor on Charles Lyell's <u>Principles of Geology</u>]: Philosophical Magazine, new series, vol. 8, pp. 215-219.

William Conybeare, the geologist who named and described the Carboniferous System, said that the uniformitarian hypothesis advanced by Charles Lyell was ". . . one of the most gratuitous and unsupported assertions ever hazarded" (page 218).

12 Hooykaas, R., 1970, Catastrophism in geology, its scientific character in relation to actualism and uniformitarianism: Amsterdam, North-Holland, 50 pp.

A renowned historian of science tells the history of catastrophism in a different fashion than geology books.

> The history of geology has often been expounded, in the fashion of a fairy tale, as a battle between good and evil. Neptunism is black, Plutonism white; Catastrophism is black, Uniformitarianism white. In the 18th century darkness reigned until, through Hutton, suddenly all became light. In the beginning of the 19th century Cuvier, Buckland, c.s. fell back again upon deluges and catastrophes, until Lyell dispelled the clouds and definitively established uniformitarian orthodoxy.

> Catastrophists are accused of giving free play to their phantasy, of rashly resorting to extraordinary events and supernatural causes, and of mixing up independent geological research with metaphysical beliefs.
>
> In this paper we will listen to the other side too. And the conclusion will be that, though there have been catastrophists who answer to the description just given, uniformitarians could be as metaphysical and perhaps even more dogmatical than their opponents, and that, quite apart from the resulting theoretical _system_, at least the _method_ of the Catastrophists was a legitimate one. (page 5)

13 Rudwick, M. J. S., 1971, Uniformity and progression: reflections on the structure of geological theory in the age of Lyell, _in_ Roller, D. H. D., ed., Perspectives in the history of science and technology: Norman, University of Oklahoma Press, pp. 209-227.

The historian of geology, Martin Rudwick, tells the history of the uniformitarian-catastrophist debate differently than popular geology textbooks. Rather than being a science dominated by fanciful and speculative theories, catastrophist geology of the early nineteenth century was a "synthesis of great scientific power and sophistication." Charles Lyell was the uniformitarian theorist, whereas the catastrophists were the scientific empiricists.

14 Grinnell, G., 1976, The origins of modern geological theory: Kronos, vol. 1, no. 4, pp. 68-76.

Undaunted by Scrope's failure, the young whig lawyer Charles Lyell now tried his hand at destroying the geological foundation of monarchical theory. In his Principles of Geology he took a much more subtle line than had Scrope. In the 100-page introduction to the Principles, Lyell argued not so much that the diluvial theory was wrong, as that it was mythological and impeded the "progress" of geology. In the first volume he went on at great length concerning the forces of erosion and the effects of volcanic uplift in what was a brilliant avoidance of all evidence of catastrophism. It was just what the moderates were looking for. They rallied around Lyell and elected him secretary first, and then president of the Geological Society.

In this day and age when geology is far removed from religion and politics and when political issues are settled by election rather than at meetings of geological societies, it is difficult for us to understand the extent to which the social shift in world view which took place not only in geology but in astronomy and in natural history was related to the Great Reform movement of 1832. All were part of the far more general shift in world view from paternalism to liberalism, but the persons responsible for engineering this shift were very conscious of what they were doing. "It is a great treat to have taught our section-hunting quarry men, that two thick volumes may be written on geology without once using the word "stratum," Scrope wrote to Lyell on September 29, 1832, after Lyell's second volume appeared. "If anyone had said so five years back, how he would have been scoffed at." Just as the conservatives had refused a hearing to the Huttonian camp earlier, now the liberals pulled the same tactics when they got into power. The stronghold of catastrophism lay in a stratigraphy where unconformity and nonconformities, to say nothing of massive conglomerates, told of wide-ranging geological disasters in the past. Lyell, like Scrope before him, simply suppressed the evidence which did not fit in with his doctrines, and once he was voted into power, the catastrophists found it increasingly difficult to publish their research.

The liberal takeover of the geological society and the suppression of evidence favoring the catastrophist position did not come about overnight. Rather there was a slow assimilation of catastrophist data until there was virtually nothing left to the theory as a whole. When, in 1839, Louis Agassiz attempted to argue in favor of catastrophism with his theory of ice ages, the uniformitarians simply adopted all his evidence, but reinterpreted it in uniformitarian terms. Thus the data did not change, but the Gestalt by which that data was organized and given coherence was transformed from catastrophism to uniformitarianism, just as the social structure of England was changed from Tory Paternalism in which sovereignty descended from God down to the King, to the new liberalism in which sovereignty ascended up from the people through Parliament to its ministers.

Ironically enough, the political battle which underlay the catastrophist-uniformitarian debate of 1832 is now long over, but owing to the paradigmization of science, the uniformitarian Gestalt is still assiduously cultivated at universities and in professional geological societies. The "cause" for which Babbage, Scrope, and Lyell were fighting is now long since over, and we should feel free to look again at the geological evidence itself, which, if the truth be told, provides ample evidence for catastrophism, as it always has. (pages 73-75)

The Failings of Uniformitarianism

15 Shea, J. H., 1982, Twelve fallacies of uniformitarianism: Geology, vol. 10, pp. 455-460.

> Ask almost any geologist and you will be told that uniformitarianism is <u>the</u> basic principle of geology....Beyond that, however, inquiry about this principle is likely to elicit an astonishing array of vague catch-phrases, half-truths, and outright fallacies. Similarly, geological books, journal articles, textbooks at all levels, dictionaries, and encyclopedias are riddled with misconceptions and fallacious statements of and about uniformitarianism. (page 455)

16 Gould, S. J., 1965, Is uniformitarianism necessary?" American Journal of Science, vol. 263, pp. 223-228.

> Uniformitariansm is a dual concept. Substantive uniformitarianism (a testable theory of geologic change postulating uniformity of process rates or material conditions) is false and stifling to hypothesis formation. Methodological uniformitarianism (a procedural principle asserting spatial and temporal invariance of natural laws) belongs to the definition of science and is not unique to geology. (page 223)

17 Albritton, C. C., Jr., 1963, Philosophy of geology: a selected bibliography and index: <u>in</u> Albritton, C. C., Jr., ed., The fabric of geology: Stanford, Freeman Cooper & Co., pp. 262-363.

History and Philosophy

"The present is the key to the past," we say. The principle of uniformity, to which this adage refers, is held by many to be the foundation stone of geology. With regard to the validity of the principle, however, the range of opinion is amazing. Most modern historians of science seem to agree that Lyell's famous principle was an a-historic device, which was discarded after evolutionism became popular in the nineteenth century. Most modern philosophers of science seem to feel either that the principle is too vague to be useful, or that it is an unwarranted and unnecessary assumption. Geologists and other scientists have such varied opinions on the matter that it would be impossible, without a vote, to say which view prevails. Surely the principles of uniformity needs the critical attention of geologists. (page 263)

18 Goodman, N., 1967, Uniformity and simplicity, in Albritton, C. C., ed., Uniformity and simplicity, a symposium on the principle of the uniformity of nature: Geological Society of America Special Paper 89, pp. 93-99.

A philosopher of science argues that the geologist's principle of uniformity dissolves into a principle of simplicity.

19 Twain, Mark, 1874, Life on the Mississippi: New York, Harper & Brothers, edition of 1951.

Mark Twain chides the method of uniformitarian geologists.

> Now, if I wanted to be one of those ponderous scientific people, and "let on" to prove what had occurred in the remote past by what had occurred in a given time in the recent past, or what will occur in the far future by what has occurred in late years, what an opportunity is here! Geology never had such a chance, nor such exact data to argue from! Nor "development of species," either! Glacial epochs are great things, but they are vague--vague. Please observe:
>
> In the space of one hundred and seventy-six years the Lower Mississippi has shortened itself two hundred and forty-two miles. That is an average of a trifle over one mile and a third per year. Therefore, any calm person, who is not blind or idiotic, can see that in the Old Oolitic Silurian Period, just a million years ago next November, the Lower Mississippi River was upward of one million three hundred thousand miles long, and stuck out over the Gulf of Mexico like a fishing-rod. And by the same token any person can see that seven hundred and forty-two years from now the Lower Mississippi will be only a mile and three-quarters long, and Cairo and New Orleans will have joined their streets together, and be plodding comfortably along under a single mayor and a mutual board of aldermen. There is something fascinating about science. One gets such wholesale returns of conjecture out of such a trifling investment of fact. (pages 155, 156)

The Opposition to Catastrophism

20 Shea, J. H., 1982, Uniformitarianism and sedimentology: Journal of Sedimentary Petrology, vol. 52, pp. 701, 702.

Shea comments on the idea that the geologist's doctrine of uniformitarianism opposes catastrophic events.

> This dogma is obviously a product of the history of geology. Unfortunately, we have overreacted to that history and have adopted an excessively gradualist view of earth history, refusing in many cases to consider catastrophic events (such as the Spokane flood or the impact of giant meteorites) even when the evidence clearly suggests that sudden, violent, cataclysmic events have occurred. This attitude is changing, however, and we need to free ourselves completely from the artificial constraints of a fallacious dogma that would preclude any possibility of natural catastrophes having occurred even if the postulated catastrophes are perfectly rational and supported by strong evidence. (page 702)

21 Friedman, G. M., and Sanders, J. E., 1978, Principles of Sedimentology: New York, John Wiley, 792 pp.

> Late in the nineteenth century and early in the twentieth century, a school of thought, which we shall refer to as "gradualism," became established in the United States. This school included staunch uniformitarians who opposed any idea that hinted of catastrophic activities. Such ideas were stigmatized under the heading of "Catastrophism." (pages 238-239)

History and Philosophy

22 Spencer, E. W., 1962, Basic concepts of historical geology: New York, Crowell, 504 p.

> The very word 'catastrophism' has a bad connotation for most earth scientists. They are so adamant in refuting the more extreme views of the catastrophists that they prefer not to use the word at all. (page 51)

23 Brenner, R. L. and Davies, D. K., 1973, Storm-generated coquinoid sandstone: genesis of high-energy marine sediments from the Upper Jurassic of Wyoming and Montana: Geological Society of America Bulletin, vol. 84, pp. 1685-1698.

Shell-dominated sedimentary layers (coquina) are thought to have been winnowed by water currents from dispersed shells in mud during intense storms.

> In general, analyses of ancient environments reflect the pervasive opinion that sediment formation and dispersal owe their genesis to the operation of "normal" processes. Application of what might be termed <u>noncatastrophic uniformitarianism</u> has dominated sedimentologic thought. As a result, storms and their effects have found no place in most studies of ancient sediments. (page 1685)

24 Baker, V. R., 1978, The Spokane flood controversy and the martian outflow channels: Science, vol. 202, pp. 1249-1256.

Those who uphold a strictly uniformitarian interpretation for the earth's geologic features would do well to read this somewhat tragic account of how a man's catastrophic flood hypothesis took over 40 years to be accepted by the scientific community. Baker says:

> In a series of papers published between 1923 and 1932, J Harlen Bretz described an enormous plexus of proglacial stream channels eroded into the loess and basalt of the Columbia Plateau, eastern Washington. He argued that this region, which he called the Channeled Scabland, was the product of a cataclysmic flood, which he called the Spokane flood. Considering the nature and vehemence of the opposition to his hypothesis, which was considered outrageous, its eventual scientific verification constitutes one of the most fascinating episodes in the history of modern science. (page 1249. Copyright 1978 by the American Association for the Advancement of Science.)

Baker's paleohydraulic reconstruction indicates that the discharge of the flood was as great as 21.3×10^6 cubic meters per second. Maximum flow velocity is estimated to be 30 meters per second and the water depths may have exceeded 100 m. Bretz (1969, reference 174) has summarized the geologic evidence for the Spokane flood.

25 Olson, E. C., 1969, Introduction to J Harlen Bretz's paper on "The Lake Missoula floods and the Channeled Scabland": Journal of Geology, vol. 77, pp. 503, 504.

This paper introduces the summary report of Bretz (1969, reference 174). Olson says: "During its not always calm history, the story of the development of the Channeled Scabland was thought by some to have brushed beyond the dividing line in flaunting catastrophe too vividly in the face of the uniformity that had lent scientific dignity to interpretation of the history of the earth." (page 503)

26 Anonymous, 1980, GSA medals and awards: Geological Society of America News & Information, vol. 2, no. 3, p. 40.

This news article describes the honoring of Dr. J Harlen Bretz with the Geological Society of America's highest award. He was recognized for his discovery in the 1920's that "the unique features of the Scabland topography of the Columbia plateau in southeastern Washington could only have been formed by a sudden catastrophic flood of a magnitude previously unknown in geologic science." In his acceptance speech Dr. Bretz said, "...I can be credited with reviving and demystifying legendary Catastrophism and challenging a too rigorous Uniformitarianism."

27 Hsü K. J., 1983, **Actualistic catastrophism, address of the retiring President of the International Association of Sedimentologists: Sedimentology, vol. 30, pp. 3-9.**

Sedimentologists have denied the action of catastrophic sedimentary events and need to accept a working philosophy of "actualistic catastrophism."

Chapter 3

COSMIC CATASTROPHES

A large number of meteorites were discarded by European museums in the 18th century because the French Academy of Sciences could not tolerate "superstitions" that rocks fall out of the sky. Similarly, Thomas Jefferson, scientist and U. S. President, said, "It is easier to believe that two Yankee Professors would lie, than that stones would fall from Heaven." Today, we know that rocks do fall out of the sky. Impact craters (e.g., Meteor Crater, Arizona) prove that large objects have hit the earth. But, have collisions with large asteroids or comets occurred? This chapter explores the literature dealing with catastrophic impact events and their effects on earth. The chapter describes the intriguing Tunguska event (1908) over Siberia (references 28 through 31), discusses the probability of asteroid collision (references 32 through 35), reviews the physics of impact with rock (reference 36), relates ancient impact evidences (references 37 through 44), describes theories on the origin of tektites (references 45 through 47), explores the effects of enormous impact events (references 48 through 52), and speculates on the effects of supernovae near earth (references 53 and 55).

The Tunguska Explosion of 1908

28 Brazo, M. W., and Austin, S. A., 1982, The Tunguska explosion of 1908: Origins, vol. 9, pp. 82-93.

The 1908 atmospheric explosion over Siberia has generated enormous speculation and controversy as to its origin. The theories offered range from the realm of science (a meteorite, comet, or nuclear explosion) to the realm of science fiction (a black hole, anti-matter rock, or an alien spacecraft). The best explanation appears to be an explosion of a small comet.

> Above central Siberia on June 30, 1908, at approximately 7:17 AM local time, a small comet entered the atmosphere from behind the sun and moved in a southeast to northwest direction. The comet was composed of about 30,000 tons of water, methane, and ammonia ice with traces of silicates and iron oxides. Penetrating the atmosphere at approximately 60 km/sec (130,000 mph), the object created an intense shock wave which wrapped tightly around its nose. As it descended that sunny morning, its nucleus exploded (possibly 3 times) approximately 8 km above the Earth's surface. A huge black cloud immediately appeared following the explosion which released 10^{23} ergs of energy. A heat wave with a temperature of approximately 16.6 million degrees Celsius at the focus was generated that had a tree-scorching effect for a radius of 15 km. The heat wave was followed by air shock waves which disfigured or toppled 80 million trees occupying approximately 8000 km of

Siberian taiga (a radius of 30 km), and initiated a seismic wave of Richter magnitude 5, but, to our astonishment, left no crater. The dust from the tail of the comet moved away from the sun and provided anomalously bright night sky in Europe and parts of Western Russia. No trace of the comet itself was found except for tiny magnetite and silicate globules. The principal consequences were fear and awe among the inhabitants of the region, and the physical damage from the explosion. Fortunately, no human life was lost, though more than a thousand reindeer were destroyed. (pages 91 and 92)

29 Ben-Menahem, A., 1975, Source parameters of the Siberian explosion of June 30, 1908, from analysis and synthesis of seismic signals at four stations: Physics of Earth and Planetary Interiors, vol. 11, pp. 1-35.

The explosion of a comet or meteorite in the atmosphere over Siberia in 1908 released an estimated 5×10^{23} ergs (12.5 megatons equivalent) of energy, about 1/100,000 of which was seismic waves (Richter magnitude 5.0). As the object penetrated the atmosphere, it had a gaseous aura, pale-blue in color with a disappearing trail of thermally ionized air. The heat from the spherical source was felt at 70 kilometers distance, the explosion at a height of 8 kilometers above the surface could be seen at a distance of 500 kilometers and heard over a distance of 1270 kilometers. Trees were toppled inside a radius of 25 kilometers, but trees on high ground were uprooted at a distance of up to 50 kilometers. No reliable reports of felt ground motion have been obtained at appreciable distance from the explosion.

30 Krinov, E. L., 1966, Giant meteorites: New York, Pergamon Press, 397 p.

Krinov, a Russian scientist, has published an extensive investigation of the Tunguska event of Siberia of 1908. Included in the Russian translation is the eyewitness account of S. B. Semenov who was 60 kilometers south of the blast site in the village of Vanovara where he was blown off a porch by the air wave.

> ... I was sitting in the porch of the house at the trading station of Vanovara at breakfast time... when suddenly in the north ... the sky was split in two and high above the forest the whole northern part of the sky appeared to be covered with fire. At that moment I felt great heat as if my shirt had caught fire; this heat came from the north side. I wanted to pull off my shirt and throw it away, but at that moment there was a bang in the sky, and a mighty crash was heard. I was thrown to the ground about three sajenes [about 7 meters] away from the porch and for a moment I lost consciousness... The crash was followed by noise like stones falling from the sky, or guns firing. The earth

trembled, and when I lay on the ground I covered my head because I was afraid that stones might hit it. (pages 147 and 148)

31 Stanyukovich, K. P., and Bronshten, I. A., 1961, Velocity and energy of the Tunguska meteorite: Doklady Akademii Nauk SSSR, Earth Sciences Sections, vol. 140, pp. 1053-1055.

This paper is an energy estimate of the mysterious event in Siberia on June 30, 1908 which knocked down and "toasted" a forest 30 kilometers in diameter inflicting lesser damage over a much broader area. The authors believe that the destruction was caused by an above-ground explosion of a meteorite. Other Russian scientists have suggested the explosion of a small comet. Energy estimates vary from 10^{20} to 10^{23} ergs with the authors suggesting the higher value. No meteorite fragments have been found and no distinct crater formed.

Probability of Asteroid Impact

32 Wetherill, G. W., 1979, Apollo objects: Scientific American, vol. 240, no. 3, pp. 54-65.

The term "Apollo object" is given to any asteroid possessing an earth-crossing orbit. Wetherill believes that more than 750 Apollo objects greater than 1 kilometer diameter exist and that there is one chance in 250,000 that such an object will strike the earth this year. This is stated by Wetherill in a different way; the frequency of collision with these large objects would be about 4 times per million years. Wetherill's estimates agree with those of Shoemaker et al. (1979, reference 33). Wetherill says:

> In 1937 a body about a kilometer in diameter, later named Hermes, passed within 800,000 kilometers of the earth, no more than twice the distance of the moon. It has not been seen again. About once in every century a similar object can be expected to travel past the earth at less than the lunar distance. And once in every 250,000 years, on the average, the earth and such a body will collide. The impact of the collision will release energy equivalent to 10,000 10-megaton hydrogen bombs and will make a crater some 20 kilometers in diameter. Fortunately such catastrophes are so infrequent that none has been recorded within human history. (page 54, Copyright 1979 by Scientific American, Inc. All rights reserved.)

33 Shoemaker, E. M., Williams, J. G., Helin, E. F., and Wolfe, R. F., 1979, Earth-crossing asteroids: orbital classes, collision rates with earth, and origin, in Gehrels, T., ed., Asteroids: Tucson, University of Arizona Press, pp. 253-282.

The authors estimate that there are presently about 1300 asteroids with diameters greater than 1 kilometer possessing earth-crossing orbits. About 700 of these asteroids are considered to belong to the Apollos class which has the greatest probability of collision with earth. The collision rate with earth-crossing asteroids larger than 1 kilometer diameter is calculated to be about one chance in 286,000 this year, or about 3.5 per million years. If we extrapolate this rate over the 4.5-billion-year age assumed for the earth, about 16,000 asteroids larger than 1 kilometer should have impacted the earth; about 2100 collisions should have occurred since the beginning of the Cambrian Period (age estimated by radiometric dating at 600 million years). Because a 1 kilometer diameter asteroid should produce a 22-kilometer-diameter crater (Dence et al., 1977, reference 36), the continents, which comprise about 1/3 of the surface of the globe, should contain about 700 Phanerozoic impact craters greater than 22 kilometers diameter. Grieve and Robertson

(1979, reference 37) catalog only 21 probable Phanerozoic impact craters larger than 22-kilometer-diameter from the continents. It appears unlikely that the remaining 680 unaccounted for craters remain unidentified, buried or were removed by erosion. Where are they? Shoemaker et al. suggest that the collision frequency was less than at present over most of geologic time, and dispute the idea of exponentially declining abundance of large earth-crossing asteroids through time as has been supposed in some evolutionary models for the origin of the solar system. They suppose there is a mechanism for continually introducing new earth-crossing asteroids through time. Another possibility is that geologists have overestimated the duration of Phanerozoic time.

Wetherill (1979, reference 32) agrees with the collision rate estimate of Shoemaker et al. and confirms the abundance of Apollo objects. He supposes that there are at least 750 Apollo objects with diameters in excess of 1 kilometer.

34 Chapman, C. R., and Davis, D. R., 1975, Asteroid collisional evolution: evidence for a much larger early population: Science, vol. 190, pp. 553-556.

The present population of earth-crossing asteroids is a remnant of a vastly larger early population. This conclusion is contrary to that of Wetherill (1979, reference 32) and Shoemaker et al. (1979, reference 33).

35 Clube, V., and Napier, W. M., 1982, The cosmic serpent: New York, Universe Books, p. 299.

Two astronomers claim that impacts from comets, asteroids, and meteoric bombardments caused devastation on a global scale. Cosmic catastrophes even occurred in historical times and gave rise to myths and legends. Extrapolating from the known Apollos asteroids, the conclusion is reached that during the last 5,000 years the earth should have been impacted by 50 objects with energy range of 1 to 100 megatons, five objects with energy range 100 to 1,000 megatons, and a 50% chance of one object with energy range of 1,000 to 10,000 megatons.

Physics of Impact with Rock

36 Dence, M. R., Grieve, R. A. F., and Robertson, P. B., 1977, Terrestrial impact structures: principal characteristics and energy considerations, in Roddy, D. J., Pepin, R. O., and Merrill, R. B., eds., Impact and explosion cratering: New York, Pergamon Press, pp. 247-275.

The paper describes the types and structures of terrestrial impact craters, then uses energy scaling from nuclear and chemical explosions to derive equations for estimating kinetic energies for large terrestrial impact craters. For terrestrial craters in crystalline rock the following equations are given:

$$D_R = 1.96 \times 10^{-5} E^{1/3.4}$$

for $D_R \geq 2.4$ kilometers

and

$$D_E = 9.7 \times 10^{-5} E^{1/4}$$

for $D_E \geq 6.4$ kilometers

where D_R is the rim diameter in kilometers of the resultant crater, D_E is the diameter in kilometers of the transient excavation cavity, and E is the kinetic energy in joules of the asteroid. The Sudbury structure of Ontario, Canada, is estimated to have required an asteroid of $E = 2.1 \times 10^{23}$ joules (2.1×10^{30} ergs or the equivalent of 5×10^7 megatons of TNT).

Other terrestrial effects of large asteroid or comet impact are described by Clube and Napier (1982, reference 48).

Ancient Impact Evidences

37 Grieve, R. A. F. and Robinson, P. B., 1979, The terrestrial cratering record--1. Current status of observations: Icarus, vol. 38, pp. 212-229.

This catalog lists proven impact structures (with associated meteorites) and probably impact structures (with associated shock metamorphic products). The nine largest probable impact structures are listed in the following table.

EARTH'S LARGEST ASTEROID IMPACT CRATERS

Name/Location	Crater Diameter (km)	Impact Energy* (ergs)
Sudbury/Ontario, Canada	140	2.1×10^{30}
Vredefort/South Africa	140	2.1×10^{30}
Popigai/Taymyr, USSR	100	6.7×10^{29}
Puchezh-Katunki/Russian SFSR, USSR	80	3.1×10^{29}
Manicouagan/Quebec, Canada	70	2.0×10^{29}
Siljan/Sweden	52	7.2×10^{28}
Kara/Yamal-Nenets, USSR	50	6.3×10^{28}
Charlevoix/Quebec, Canada	46	4.7×10^{28}
Araguainha Dome/Brazil	40	3.0×10^{28}

*Impact energies estimated using scaling equation of Dence et al. (1977, reference 36)

38 Dietz, R. S., 1964, Sudbury structure as an astrobleme: Journal of Geology, vol. 72, pp. 412-434.

This paper presents evidence for asteroid impact origin of the Sudbury structure in northern Ontario, Canada. Shock metamorphic products include shock breccia, shatter cones, and impact-melted igneous rocks. Sudbury produces 75 percent of the western world's nickel, and accounts for about 20 percent of Canada's mineral wealth. Dence et al. (1977, reference 36) estimates that the 140-kilometer-diameter crater was produced by an asteroid having 2.1×10^{30} ergs of kinetic energy (the equivalent of 5×10^7 megatons of TNT).

39 Masaytis, V. L., Mikhaylou, M. V., and Selivanovskaya, T. V., 1972, The Popigay meteorite crater: International Geology Review, vol. 14, pp. 327-331.

The Popigay crater is a 100 kilometer diameter structure on the Central Siberian Platform possessing shatter cone shock metamorphism and a large quantity of impact breccia. The scaling equation of Dence et al. (1977, reference 36) indicates that the impact energy is about 6.7×10^{29} ergs.

40 Zeylik, B. S., and Seytmuratova, E. Y., 1975, Giant meteorite impact structure in central Kazakhstan and its magma- and ore-controlling significance: Doklady Akademii Nauk SSSR, Earth Science Sections, vol. 218, pp. 26-29.

This paper is a description of a 700-kilometer-diameter circular structure in Kazakhstan in south-central Russia (the Ishim structure). It is a possible asteroid impact feature. The shock metamorphic evidences are not conclusive and we await a more thorough study.

41 Klasner, J. S., and Schulz, K. J., 1982, Concentrically zoned pattern in the Bouguer gravity anomaly map of northeastern North America: Geology, vol. 10, pp. 537-541.

A 2,800-kilometer-diameter concentrically zoned gravity high occurs in northeastern North America having its center located about 400 kilometers north of Lake Superior. Precambrian foldbelts tend to parallel and mimic the concentric pattern which is the possible vestige of an enormous asteroid impact structure.

42 Norman, J., Price, N., and Chuckuru-Ike, M., 1977, Astrons—the Earth's oldest scars?: New Scientist, vol. 73, pp. 689-692.

Large-scale, arcuate crustal features (curves in the coasts of China and West Africa, curved mountainous coast of eastern Australia, etc.) are possible remnants of 3000-kilometer-diameter scars produced by asteroid impacts. These enormous scars are believed to have formed after collisions with 300-kilometer-diameter asteroids having kinetic energies of approximately 10^{35} ergs.

13 Sawatsky, H. B., 1975, Astroblemes in Williston basin: American Association of Petroleum Geologists Bulletin, vol. 59, pp. 694-710.

Three structures within the Williston basin (Saskatchewan, Manitoba and North Dakota) are thought to be buried impact structures. Two of the structures have produced oil. Sawatsky argues that the third should be more thoroughly drilled.

4 Alvarez, L. W., Alvarez, W., Asara, F., and Michel, H. V., 1980, Extraterrestrial cause for the Cretaceous-Tertiary extinctions: Science, vol. 208, pp. 1095-1108.

Iridium, one of the platinum metals, is depleted in the earth's crust relative to its cosmic abundance. Sedimentary layers rich in iridium are considered to contain extraterrestrial material. Alvarez et al. correlate three iridium-rich limestones from Italy, Denmark and New Zealand and postulate that the collision of a large asteroid caused the extinction of the dinousaurs, flying reptiles, and various marine invertebrates and vertebrates at the Cretaceous-Tertiary boundary. Assuming that the iridium-rich layer contains impact ejecta circulated on a global scale, the stratospheric dust would have obscurred the sun's light for a short time with severe biological consequences. The asteroid is thought to have had a diameter of 10±4 kilometers, struck the earth at about 25 kilometers per second (55,000 miles per hour), and delivered 4×10^{30} ergs (equivalent to 10^8 megatons of TNT) of kinetic energy to the atmosphere and geoid. The mass of the asteroid is postulated to have been about 1.2×10^{18} grams (1.3×10^{12} tons). The ejecta mass is estimated to have been 6.9×10^{19} grams (7.6×10^{13} tons) and would have thrown about 3.5×10^4 cubic kilometers (8.4×10^3 cubic miles) of material into the air (about two thousand times larger than the 1883 Krakatoa eruption), 22 percent of which is assumed to have circulated through the stratosphere. The impact of a 10-kilometer asteroid would produce a crater 200 kilometers in diameter, larger than Sudbury and Vredefort, the two largest probable impact craters yet identified (see

Grieve and Robertson, 1979, reference 37). Because no appropriate crater is known on the continents, it appears that the asteroid struck the ocean. If the asteroid impacted the ocean, a series of huge sea waves (tsunamis) would have been created (see Gault et al., 1979, reference 186), a subject not discussed by the authors. Catastrophic hydraulic sedimentary processes should be recognizable at the Cretaceous-Tertiary boundary.

Tektites

45 Smith, P. J., 1982, The origin of tektites--settled at last?: Nature, vol. 300, pp. 217,218.

Tektites are drop-like or button-like glassy stones that exhibit every appearance of once being liquid droplets of rock solidified in flight through the atmosphere. The glassy stones are found primarily in four strewn fields in Europe, Australia, North America, and Africa. Most scientists believe that tektites are of terrestrial origin with meteorite impact serving as the agent which ejected and melted the stones. A few scientists suggest that tektites were ejected from the moon by meteorite impact or lunar volcanic eruption.

46 O'Keefe, J. A., 1978, The tektite problem: Scientific American, vol. 239, no. 2, pp. 116-125.

The moon is littered with rocks of tektite composition and the author believes that tektites on earth were ejected from the moon by intense lunar volcanism melting upon entry into the earth's atmosphere.

47 Ganapathy, R., 1982, Evidence for a major meteorite impact on the earth 34 million years ago: implication for Eocene extinctions: Science, vol. 216, pp. 885-886.

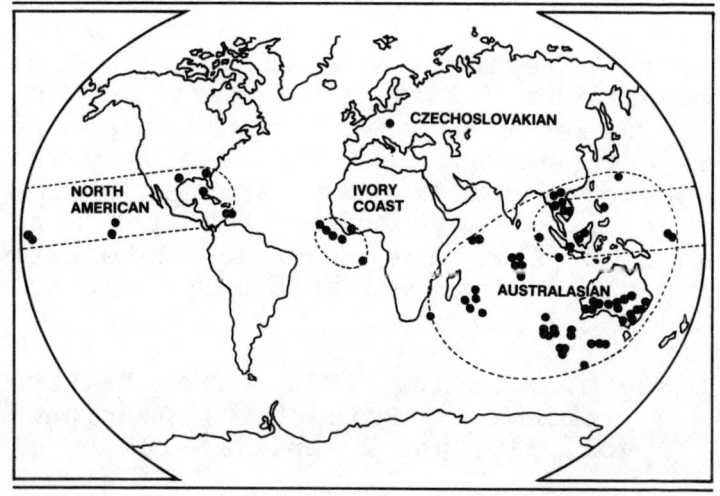

Figure 1

Location map showing where tektites and microtektites have been found. The small glassy rock fragments are believed to have formed by melting of ejecta after impact of extraterrestrial bodies. Four tektite strewn fields are indicated.

The North American microtektite field contains billions of tons of fine, glassy rock fragments in Eocene strata and is believed to have been dispersed by the impact of a 3-kilometer-diameter meteorite. At least 7 percent of the surface of the globe was dusted with small fragments which have been found in Texas, Georgia, the Caribbean Sea, and the Gulf of Mexico.

Effects of Impact

48 Clube, S. V. M., and Napier, W. M., 1982, Close encounters with a million comets: New Scientist, vol. 95, pp. 148-151.

The authors believe that the earth has been struck repeatedly by large comets. A comet 10 kilometers in diameter would cause phenomenal destruction.

> If such a comet hit land it would create a blast wave that would immediately kill off any life over a hemisphere. The air temperature would be about 500°C and windspeed about 2500 km/h at 2000 km from the site of a large impact. Hot ash streaming overhead would add to the incineration and a pall of dust would spread globally, cutting out sunlight and probably taking months to settle. Nitric oxides in the fireball would have destroyed atmospheric ozone and after the dust had settled, the Earth's surface would be exposed to ultraviolet light of germicidal intensity. Earthquakes would be global and extremely intense, with ground waves typically 10 metres high.
>
> If a comet hit the ocean it would generate waves kilometres high near the epicentre, dropping to say 0.5 km height at 1000 km but rearing up again on entering a continental shelf and running on to land. The internal currents in the Earth's core would be strongly perturbed, which would disturb its magnetic field, and so magnetic disturbances would be correlated with mass extinctions of all forms of life. But perhaps most significant, the slow viscous driving motion governing continental drift will

be violently disturbed leading to rapid plate movements, the opening of cracks 10-100 km wide in the Earth's crust, with rapid mountain building, worldwide eruptions of volcanoes (vulcanism) and so on. The Earth will eventually settle into a new, relatively undisturbed configuration but one that is biologically and geophysically very different from what went before. (pages 149, 150. Copyright New Science Publications. This first appeared in New Scientist, London, the weekly review of science and technology.)

49 Napier, W. M., and Clube, S. V. M., 1979, A theory of terrestrial catastrophism: Nature, vol. 282, pp. 455-459.

The global geologic consequences of a large asteroid collision with earth include:

(1) formation of a crater (if impact occurs on land),
(2) generation of giant sea waves (if impact occurs in ocean
(3) production of enormous overpressure in the atmosphere due to the blast,
(4) ejection of large amounts of debris including dust in the stratosphere,
(5) initiation of global cooling, perhaps even an ice age,
(6) release of NO gas and destruction of ozone layer with accompanying influx of ultraviolet light,

(7) change in rotational period of earth,
(8) possible reversal of earth's magnetic field, and
(9) mass extinction of organisms.

50 Dachille, F., 1963, Axis changes in the earth from large meteorite collisions: Nature, vol. 198, p. 176.

Although asteroid collisions would make only slight changes in the axis and rotation rate, the geologic effects could be significant. The theoretical study indicates that a 32-kilometer-diameter asteroid could deflect the earth's axis by an angle of 2 seconds and change the earth's rotational velocity by 0.001 percent. Effects induced would include worldwide sea level change.

51 Hsü, K. J., 1980, Terrestrial catastrophe caused by cometary impact at the end of Cretaceous: Nature, vol. 285, pp. 201-203.

Evidence is presented indicating that the extinction, at the end of the Cretaceous of large terrestrial animals was caused by atmospheric heating during a cometary impact and that the extinction of calcareous marine plankton was a consequence of poisoning by cyanide released by the fallen comet and of a catastrophic rise in calcite-compensation depth in the oceans after the detoxification of the cyanide. (page 201)

52 Calame, O., and Mulholland, J. D., 1978, Lunar crater Giordano Bruno: A.D. 1178 impact observations consistent with laser ranging results: Science, vol. 199, pp. 875-877.

J. B. Hartung has proposed that the chronicles of Gervase of Canterbury include striking eyewitness accounts from "reliable" observers of an asteroid impact on the lunar surface on the evening of 18 June 1178, Julian calendar. Calame and Mulholland test the idea that the 20-kilometer-diameter crater Giordano Bruno formed by the impact 800 years ago. The event would have been impressive when viewed from earth and is consistent with present free oscillations in the rotational motion of the moon.

Cosmic Catastrophes

Figure 2

Impact craters on the far side of the Moon. Most of the Moon's surface is so densely cratered that newer craters serve only to obliterate older craters. No uncratered surface exists. The largest lunar craters require impact of objects which had kinetic energies sometimes exceeding 10^{31} ergs.

[Apollo 11 photo courtesy of NASA]

Effects of Supernovae

53 Clark, D. H., McCrea, W. H., Stephenson, F. R., 1977, Frequency of nearby supernovae and climatic and biological catatstrophes: Nature, vol. 265, pp. 318, 319.

At least five historically verified supernovae have been observed in the last 1000 years in the near-side of our Galaxy. The frequency of supernovae in our Galaxy is estimated to be one per 100 years. The authors describe the effects of a supernovae occurring close to earth at a distance of 10 parsecs (32.6 light-years). The earth's atmosphere would receive about a million ergs per square centimeter as ionizing radiation which would generate oxides of nitrogen which would catalytically destroy ozone. With a depleted ozone layer solar ultraviolet radiation would be lethal to many living things. In addition, a hundredfold increase in cosmic ray intensity would produce a thirtyfold increase in the mean level of radioactivity at the earth's surface. The authors believe that such an event may explain the extinction of ancient life.

54 Reid, G. C., McAfee, J. R., and Crutzen, P. J., 1978, Effects of intense stratospheric ionizing events: Nature, vol. 275, pp. 489-492.

Protons emitted from a large solar flare or supernova would cause ionization of gases in the stratosphere. Of most importance is the production of nitrogen oxides and the consumption of large amounts of ozone. The result at the earth's surface would be high levels of damaging ultraviolet radiation, cooler global temperatures, and reduced photosynthetic activity. The authors believe that intense ionizing events have occurred on earth in the past.

55 Kloosterman, J. B., Powell, J. E., Bogoslovsky, V., Koren, K., Warth, M., Russell, D. A., 1978, Radioactive fossil bones: Catastrophist Geology, vol. 3, no. 1, pp. 4-11.

The search for radioactive fossil bones such as would be formed on earth due to cosmic ray bombardment by a supernova is reported. Radioactive fossil bones are common, but, so far, the evidence for cosmic ray origin is inconclusive.

Chapter 4

EXTRUSIVE AND INTRUSIVE CATASTROPHES

Fluid materials move inside of the earth and occasionally penetrate crustal rocks. We call these **intrusive** materials. Sometimes fluid substances break the earth's surface and accumulate where we can see them. We call these **extrusive** materials. The most common intrusive material is molten rock, which, inside the earth, is called **magma**. Molten rock frequently breaks the earth's surface where it becomes extrusive material such as **lava** or fragmental debris called **tephra**. Cold intrusive and extrusive processes also occur by fluidization of solids frequently forming clastic dikes and related features. This chapter reviews modern volcanic processes and structures (references 56 through 77), describes prehistoric pyroclastic and lava flow deposits (references 78 through 93), analyzes hot and cold intrusive processes (references 94 through 103), and discusses sedimentary products of catastrophic volcanism (references 104 through 108).

Volcanoes in History

56 Decker, R., and Decker, B., 1981, Volcanoes: San Francisco, W. H. Freeman, 224 pp.

One of the largest explosive eruption on historic record is the Tambora, Sumbawa, Indonesia event of 1815:

> Its giant eruption in 1815 may have exceeded the size and power of Krakatau. The explosion, followed by caldera collapse, is estimated to have produced between 30 and 150 cubic kilometers of ash and blocks. Ten thousand people were killed by the eruption and 80,000 starved in the resulting crop loss and famine. World climate may have been affected. The last eruption was before 1913. (page 222)

The largest historic lava flow, the Laki flow of Iceland in 1783, issued from a fissure 25 kilometers long:

> Its single giant eruption in 1783 produced over 12 cubic kilometers of lava, a historic record, filling two river valleys and covering more than 500 square kilometers. Stunted grass and fluorine poisoning from the accompanying volcanic gases starved and killed most of Iceland's livestock. The ensuing famine caused 10,000 deaths. (page 220)

Concerning ancient volcanic deposits, the authors say:

> Huge deposits of pyroclastic flows that cover thousands of square kilometers and are tens to hundreds of meters thick exist in Japan, New Zealand, Central America, the western United

Figure 3

Volcanic eruption of Mt. Ngauruhoe, New Zealand, in January 1974. The rising eruption cloud is surrounded at its base by a small pyroclastic flow captured in this unique photograph in the process of cascading down the slope of the volcano. Much larger pyroclastic flows, indicating much more energetic explosions, are documented by prehistoric volcanic deposits.

[Courtesy of NOAA/EDS]

States, and many other volcanic regions of the world. Some of these deposits give every indication that they were poured out in a single enormous eruption that would dwarf Krakatau. The volume in these deposits is on the order of 100 to 1000 cubic kilometers compared to the 18 cubic kilometers of Krakatau....Krakatau is probably only a small sample of what nature can deliver in the way of a volcanic cataclysm. (page 116, copyright 1981 by W. H. Freeman and Company. All rights reserved.)

57 Decker, R., and Decker, B., 1981, The eruption of Mount St. Helens: Scientific American, vol. 244, no. 3, pp. 68-80.

The violent eruption of Mount St. Helens on May 18, 1980 was one of the most closely monitored and fully documented volcanic catastrophes in history. Earthquake activity beneath the mountain and bulging of the north side of the mountain began in March. On May 18, after 140 meters of bulging had developed, a magnitude 5.1 earthquake launched two cubic kilometers of rock from the oversteepened north slope creating a fluidized avalanche of rock and ice which accelerated rapidly to a velocity of more than 250 kilometers per hour, extending down the Toutle River a distance of 21 kilometers. The release of pressure inside the volcano by the avalanche suddenly caused superheated water to flash into steam producing the explosion of magma ejecting a northward-directed slurry of gas-suspended rock fragments. The pyroclastic flow surged to speeds up to 400 kilometers per hour, had a temperature up to 300 degrees Celsius, carried fragments several tens of centimeters in diameter to distances of up to 15 kilometers, and devastated 550 square kilometers. Severe flooding and mudflows occurred subsequently downstream on the Toutle River. The May 18 eruption displaced 2.7 cubic kilometers of material and released about 1.7×10^{25} ergs (equivalent to a 400-megaton explosion) which is eight times more energy than the largest nuclear explosion.

58 Symons, G. J., ed., 1888, **The eruption of Krakatoa and subsequent phenomena: Report of the Krakatoa Committee of the Royal Society, London, Trubner, 494 pp.**

This work describes the largest well-studied terrestrial explosion in historic times, the explosion of the island volcano, Krakatoa, in the Sunda Strait, between Java and Sumatra on August 26 and 27, 1883. The sound of the explosions was heard over 1/13 of the globe (up to 4800 kilometers away). An explosion produced a tsunami which killed 36,000 people on the low shores of Java and Sumatra.

Extrusive and Intrusive Catastrophes 77

Figure 4

Satellite photo of air shock wave 43 minutes after the catastrophic explosion of Mount St. Helens on the morning of May 18, 1980. The shock wave viewed from 22,000 miles above the earth appears as a condensation cloud surrounding the volcano in southern Washington State.
[Courtesy of NOAA/EDS]

59 Self, S., and Rampino, M. R., 1981, The 1883 eruption of Krakatau: Nature, vol. 294, pp. 699-704.

The eruption of Krakatau (1883) is described in terms of modern volcanic theory. The major eruption began at 1300 hours (local time) on August 26 with the deposition of air fall pumice and pyroclastic surge beds. At 0535 on August 27 the style of eruption changed dramatically. During the next six hours, five major explosions produced an enormous eruption column extending over 40 kilometers above the volcano. The intermittent collapse of the column above the volcano produced pyroclastic flows which issued onto the ocean, producing large tsunamis and depositing ignimbrite on the sea floor. The collapse of the magma chamber and formation of the caldera is believed to have occurred after the explosive eruptions were completed. The total bulk volume of the deposits is estimated to be 18-21 cubic kilometers (9-10 cubic kilometers dense rock volume).

60 Watson, Capt. W. J., 1883, The Java disaster: Nature, vol. 29, pp. 140, 141.

Among the closest witnesses surviving the 1883 explosion of Krakatoa were those aboard the British ship, Charles Bal, bound for Hong Kong, in the

Extrusive and Intrusive Catastrophes 79

Sundra Straits when the volcano erupted. The captain's log gives a graphic account of the explosions and tsunami (evening of August 26 and day of August 27).

> The blinding fall of sand and stones, the intense blackness above and around us, broken only by the incessant glare of varied kinds of lightning and the continued explosive roar of Krakatoa, made our situation a truly awful one. At 11 PM, having stood off from the Java shore, wind strong from the southwest, the island, eleven miles west-north-west, became more visible, chains of fire appearing to ascend and descend between the sky and it, while on the southwest end there seemed to be a continued roll of balls of white fire; the wind, though strong, was hot and choking, sulphurous, with a smell like burning cinders. From midnight to 4 AM the same impenetrable darkness continuing, the roaring of Krakatoa less continuous but more explosive in sound, the sky one second intense blackness and the next a blaze of fire; mastheads and yardarms studded with electrical glows and a peculiar pinky flame coming from clouds which seemed to touch the mastheads and yardarms. At 6 AM, being able to make out the Java shore, set sail. Passed Anjer at 8:30 AM, close enough in to make out the houses, but could see no movement of any kind. At 11:15 there was a fearful explosion in the direction of Krakatoa, now over 30 miles distant. We saw a wave rush right on to Button Island, apparently sweeping right over the south part and rising half way up the north and east sides. This we saw repeated twice, but the helmsman says he saw it once before we

looked. The same waves seemed also to run right on to the Java shore. The sky rapidly covered in, by 11:30 AM we were inclosed in a darkness that might almost be felt. At the same time commenced a downpour of mud, sand, and I know not what; ship going northeast by north, seven knots under three lower topsails; put out the side lights, placed two men on the lookout forward, while mate and second mate looked out on either quarter, and one man employed washing the mud off binnacle glass. We had seen two vessels to the north and northwest of us before the sky closed in, adding much to the anxiety of our position. At noon the darkness was so intense that we had to grope about the decks, and although speaking to each other on the poop, yet could not see each other. This horrible state and downpour of mud continued until 1:30 PM, the roarings of the volcano and lightnings being something fearful. By 2 PM we could see the yards aloft, and the fall of mud ceased. Up to midnight the sky hung dark and heavy, a little sand falling at times, the roaring of the volcano very distinct although we were fully sixty-five or seventy miles northeast from it. Such darkness and time of it few would conceive, and many, I dare say, would disbelieve. The ship, from truck to waterline, is as if cemented; spars, sails, blocks, and ropes in a terrible mess; but, thank God, nobody hurt or ship damaged. On the other hand, how fares it with Anjer, Merak, and the other little villages on the Java coast?

The great tsunami caused by the explosion passed unnoticed under their ship in deep water but reached great height at the shore. The coastal villages did not fare well. Over 36,000 people were drowned by the great sea waves.

61 Yokoyama, I., 1957, Energetics in active volcanoes, 2nd paper: Tokyo University Earthquake Research Institute Bulletin, vol. 35, pp. 75-97.

Energies of historic volcanic eruptions are estimated. The largest listed is the 1815 explosion of Tambora, Sumbawa, Indonesia which may have moved about 100 cubic kilometers of rock and released about 8.4×10^{26} ergs of energy. The Tambora eruption (1815) was about 80 times as energetic as the Krakatoa eruption (1883). The Tambora explosions had the equivalent of 2000 megatons TNT or 840 of the largest hydrogen bombs.

62 Marinatos, S., and Imboden, O., 1972, Thera: key to the riddle of Minos: National Geographic, vol. 141, pp. 702-726.

Thera (Santorin), a volcanic island in the Aegean Sea, is believed to have exploded in about 1520 B.C., destroying Minoan and Minoan-related settlements in Turkey, Greece, and Crete. An earthquake is believed to have preceded the explosion which was followed by caldera collapse and a huge tsunami ("tidal wave").

Yet, colossal as it was, the explosion of Krakatoa released a mere fraction of the destructive force unleashed 3,400 years before by the eruption of Thera. In the agony of Krakatoa, a bit more than eight square miles of the island sank into the sea; Thera lost 32 square miles. To gain an adequate picture of the convulsion that shook the Aegean, I suggest that one must multiply the Krakatoa events by a factor of four.

I calculate that the tidal waves created by the eruption sped from Thera to Crete--a distance of 70 miles--in less than half an hour. Some scholars estimate their speed at more than 200 miles an hour and assign to them the enormous height of 300 feet as they piled up on the northern coast of Crete. (page 718)

63 Watkins, N. D., Sparks, R. S. J., Sigurdsson, H., Huang, T. C., Federman, A., Carey, S., and Ninkovich, D., 1978, Volume and extent of the Minoan tephra from Santorini Volcano: new evidence from deep-sea sediment cores: Nature, vol. 271, pp. 122-126.

The eruption of the island volcano Santorini (Thera) in approximately 1500 B.C. was one of the largest in history.

Analyses of tephra in abyssal piston cores collected during cruises of R/V Trident show that the Minoan eruption produced at least 28 km^3 of tephra (13 km^3 dense rock equivalent).

A layer up to 5 cm thick must have been deposited on eastern Crete. (page 122)

64 Bond, A., and Sparks, R. S. J., 1976, The Minoan eruption of Santorini, Greece: Journal of the Geological Society of London, vol. 132, pp. 1-16.

The catastrophic explosion of Santorini (Thera) Volcano in the Aegean Sea some 3,500 years ago produced huge volumes of pumice and ash, mud flow and flood deposits, and a caldera 11.5 by 8 kilometers. The explosion is believed to have had a devastating effect on the Minoan settlements on Crete 120 kilometers to the south.

65 Boekschoten, G. J., 1971, Quaternary tephra on Crete and the eruptions of the Santorin Volcano: Opera Botanica, no. 30, pp. 40-48.

The volcanic material in the uppermost soil strata on Crete suggests accumulation from a single ashfall, the Minoan eruption of Santorin (Thera).

66 Walker, G. P. L., 1980, The Taupo pumice: product of the most powerful known (ultraplinian) eruption?: Journal of Volcanology and Geothermal Research, vol. 8, pp. 69-94.

Lake Taupo, which occupies one of the world's largest calderas, is located on the North Island of New Zealand. Taupo was the center of enormous volcanic explosions which produced one of the thinnest and most extensive volcanic layers.

> The Taupo pumice is probably the most widely dispersed fall deposit currently known, representative of a proposed new class of <u>ultraplinian</u> deposits. Although the maximum thickness is only 1.8 m, the volume is 24 km^3 (calculated by a new method based on the eolian concentration of crystals), and about 80% of this has fallen at sea farther than 220 km from source. (page 69)

Wilson et al (1980, reference 201) believe the Taupo eruption occurred in 186 A.D. making it possibly the largest explosive eruption in historical times.

67 Bullard, F. M., 1962, Volcanoes in history, in theory, in eruption: Austin, University of Texas Press, 441 pp.

The world's largest historic lava flow, the Laki flow, occurred in 1783 in southern Iceland where 12 cubic kilometers of lava devastated more than 500 square kilometers.

> Preceded by eight days of severe earthquakes, the actual outbreak began on June 8, 1783, with

tremendous explosions accompanied by vast ash clouds which rained ash over a wide area, in sufficient quantities in Scotland and Norway to damage crops. On June 11, floods of lava issued from twenty-two vents along a ten-mile-long fissure and flowed in a southwesterly direction. On reaching the Skafta River, the lava filled the valley, which was four hundred to six hundred feet deep, overflowing onto the surrounding plains. Repeated outflows occurred until the middle of July, adding to the lava in the Skafta Valley, The lava flow which flooded the Skafta Valley was fifty miles long and in the lowland area spread to a maximum width of twelve to fifteen miles, with an average thickness of about one hundred feet. In the deeper points of the valley, of course, its thickness exceeded six hundred feet. On August 3, 1783, the valley of the Hverfisfljot River, a stream parallel to the Skafta was invaded and, like the Skafta, was filled, the lava overflowing the adjoining open country. This flow was forty miles long with a maximum width of seven miles. The volume of lava from the Laki Fissure is estimated to be $12 \times 10^9 m^3$. (page 251, copyright 1962 by Fred M. Bullard. Published by the University of Texas Press.)

68 Thorarinsson, S., 1970, The Lakagigar eruption of 1783: Bulletin Volcanologique, series 2, vol. 33, pp. 910-929.

The world's largest historic lava flow, the Laki flow of 1783, devastated 565 square kilometers of southern Iceland with about 12.3 cubic kilometers of molten rock (see Bullard, 1962, reference 67). Thorarinsson estimates that the maximum rate of lava eruption in the summer of 1783 may have exceeded one kilometer per day. The area, volume and rate of eruption are much less than some ancient lava flows (see Swanson et al., 1975, reference 89).

69 Dawson, J. B., 1966, Oldoinyo Lengai--an active volcano with sodium carbonate lava flows, in Tuttle, O. F., and Gittins, J., eds., Carbonatites: New York, John Wiley, 591 p.

A sodium carbonate lava flow has been observed in the volcano Oldoinyo Lengai in northern Tanganyika. As a modern example of carbonate magma, this volcano provides insight into the strangest rocks attributed to magmatic origin--the carbonatites. The book containing this paper describes carbonatites giving evidences for magmatic origin. Some geologists believe carbonatite formed from magmas derived by rapid intrusion from the mantle of the earth, an interpretation indicated by the similarities of carbonatites and kimberlites.

70 Smith, B. A., et al., 1979, The Galilean satellites and Jupiter: Voyager 2 imaging science results: Science, vol. 206, pp. 927-950.

Enormous volcanic eruptions were observed by the first two Voyager spacecraft on Io, a large moon of Jupiter, in March and July 1979. Pyroclastic debris was observed to be ejected up to 280 kilometers (174 miles) above the moon's surface in plumes up to 1000 kilometers (620 miles) wide. One plume, near its exit from the satellite's surface, had a width of 60 kilometers (37 miles). The eruptions are the most massive volcanism known. It appears to operate continuously judging from the four-month interval between the two Voyager encounters.

71 Sparks, R. S. J., Wilson, I., and Hulme, G., 1978, Theoretical modeling of the generation, movement, and emplacement of pyroclastic flows by column collapse: Journal of Geophysical Research, vol. 83, pp. 1727-1739.

The theory and mechanics of movement of pyroclastic flows are discussed technically. A pyroclastic flow is generated when the rate of expulsion of gas released at the volcanic vent decreases to the point where it cannot support the eruption column of gas and

Figure 5

Jupiter's satellite, Io, on March 4, 1979, showing active volcanic terrain including a large donut-shaped volcanic eruption plume (approximately 400 kilometer diameter) and enormous calderas. The catastrophic volcanism on Io is dominated by sulfur with eruption rates from single vents estimated at greater than 10 million kilograms per second (11,000 tons per second). See Reference 70.
[NASA Voyager photograph P-21209C]

rock fragments above the vent. The column collapses and flows radially downhill from the vent at great speed (up to 310 meters per second or 112 kilometers per hour). The flow consists of a slurry of air, volcanic gas and rock fragments frequently hotter than 600 degrees Celsius. The turbulent flow will move and deposit hot rock fragments until enough gas has been lost from the suspension for the flow to stop. Heavy rock fragments (2.5 grams per cubic centimeter) up to a few centimeters in diameter and coarse pumice clasts (1.0 grams per cubic centimeter) up to tens of centimeters in diameter frequently can be transported tens of kilometers by the flow which retains enough heat to weld or fuse fragments in the ash matrix.

72 Waters, A. C., and Fisher, R. V., 1971, Base surges and their deposits: Capelinhos and Taal volcanoes: Journal of Geophysical Research, vol. 76, pp. 5596-5614.

A "base surge" is a ringlike turbulent flow of rock, water and gas which spreads radially at hurricane velocities from shallow underwater and underground nuclear explosions. Base-surge clouds also have been observed spreading radially from some volcanic eruption columns. The material accumulated from base surges is similar to ancient pyroclastic flow deposits, and provides an analog for understanding how this <u>common</u> type of volcanic-sedimentary <u>rock</u> formed.

Volcanic Structures

73 Ninkovich, D. and Donn, W. L., 1976, Explosive Cenozoic volcanism and climatic implications: Science, vol. 194, pp. 899-906.

A caldara is the basin structure produced by the collapse of a volcanic cone into an underground chamber evacuated by explosive eruption. The magnitude of the catastrophic eruption is indicated by the caldera volume. Crater Lake in Oregon occupies the most well-known caldera which was produced by the explosion of ancient Mount Mazama a few thousand years ago. Ninkovich and Donn describe the volumes of large calderas. These are listed here in what is believed to be historical order with the most recent at the top:

Caldera name / location	Caldera volume (cubic kilometers)	Age
Katmai, Alaska	12	historic (1912)
Krakatau, Indonesia	5	historic (1883)
Tambora, Indonesia	30	historic (1815)
Taupo, New Zealand	>70	historic (186 A.D.)
Santorini, Aegean Sea	70	historic (about 1500 B.C.)
Mount Mazama, Oregon	70	prehistoric (Holocene)
Rotorua, New Zealand	>70	prehistoric (Pleistocene)
Toba, Indonesia	2000	prehistoric (Pleistocene)

The largest of these volcanic explosions (Toba) produced an ash layer 15 to 20 centimeters thick at a distance of 1000 kilometers from the volcano.

Ancient catastrophic volcanic eruptions are believed to have introduced volcanic dust into the stratosphere and altered global climate. The authors say:

> It is true, however, that some explosive volcanism in the geologic past greatly exceeded in magnitude that in the historic past. Such events, when occurring at critical times of climate evolution, might have strongly modulated the intensity of climate change. (page 906)

74 Fairbridge, R. W., 1968, Lake Toba, in Fairbridge, R. W., ed., The encyclopedia of geomorphology: New York, Reinhold, pp. 617, 618.

What is believed to be the world's largest caldera is Lake Toba on the island of Sumatra. The lake is 88 by 29 kilometers (55 by 18 miles), covers 1,130 square kilometers (420 square miles), and has a large secondary volcanic island 43 by 19 kilometers (27 by 12 miles). The eruption of the ancient volcano deposited a blanket of tephra on Sumatra covering 20,000 square kilometers up to 600 meters thick. Toba's explosion is believed by Ninkovich, Sparks, and Ledbetter (reference 81) to have produced a deep-sea ash layer in the Indian ocean at a distance of 2500 kilometers from the volcano and the resulting collapse formed a caldera with a volume of 2000 cubic kilometers.

75 Bemmelen, R. W. van, 1930, The origin of Lake Toba (north Sumatra): Proceedings of the Fourth Pacific Science Congress, Java, vol. 2A, pp. 115-124.

The enormous caldera has a volume of 2,000 cubic kilometers which agrees approximately with the amount of blown out material.

76 Kloosterman, J. B., 1973, Giant ring volcanoes on the Guiana Shield: II Congreso Latino Americano de Geologia, Caracas, 17 pp.

Three enormous ringlike structures in volcanic and plutonic rocks of Brazil, Venezuela, Guyana, Surinam and French Guiana are believed to be the eroded remnants of gigantic Proterozoic volcanoes and their calderas. The first ring volcano is 500 by 900 kilometers, the second is 300 by 350 kilometers, and the third has a diameter of 600 kilometers. For comparison, the world's largest modern volcanic pile, the Island of Hawaii, has a diameter less than 200 kilometers.

77 Bailey, R. A., Dalrymple, G. B., and Lanphere, M. A., 1976, Volcanism, structure, geochronology of Long Valley caldera, Mono County, California: Journal of Geophysical Research, vol. 81, pp. 725-744.

Long Valley caldera is an enormous elliptical depression 17 by 32 kilometers with a depth of up to 3 kilometers on the east base of the Sierra Nevada Mountains 30 kilometers south of Mono Lake, in eastern California. The explosive Pleistocene eruptions evicerated a chamber which contained 600 cubic kilometers of rhyolite magma and formed the extensive Bishop tuff. The collapse of the volcano into the magma chamber produced the elliptical caldera, having an area of 450 square kilometers.

Prehistoric Pyroclastic Deposits

78 Heiken, G., 1979, Pyroclastic flow deposits: American Scientist, vol. 67, pp. 564-571.

Pyroclastics are any rock fragments that are explosively ejected from volcanoes. Large, explosive volcanic eruptions frequently launch slurries of air, hot gas and rock fragments which, surging in excess of highway speed, overrun the land surface at great distances from the volcano accumulating "pyroclastic flow deposits" wherever they go. Ancient pyroclastic deposits are described by Heiken:

> Pyroclastics are fragments of rock that have been explosively ejected from volcanoes. Extensive plateaus consisting of thick pyroclastic deposits are scattered over the continents, evidence of short-lived catastrophic eruptions on a scale that has never been observed by man. Such eruptions would be awesome, burying hundreds to thousands of square kilometers under pyroclastic flows that move away from the vent at hurricane velocities. Eruption clouds may deposit fallout over millions of square kilometers. A safe distance from which to watch such an event might be the earth orbit of a space station. (page 564)

Individual pyroclastic flows are described by Heiken:

> Deposits range in thickness from a few tens of meters to hundreds of meters, thinning from source to terminus. Individual pyroclastic

flows may have traveled <10 km up to 120 km and cover thousands of square km. Total volumes of tephra from these short-lived catastrophic eruptions range from tens of cubic km to over a thousand. Eruption sequences must occur over a short period of time, because those consisting of several flow units--each representing an individual explosive event lasting minutes? hours? days?--rarely show evidence of any erosion or weathering between them. (page 566)

79 Alt, D. D., and Hyndman, D. W., 1978, Roadside geology of Oregon: Missoula, Mountain Press, 268 pp.

Welded ash forms when a volcano erupts a cloud of steam mixed with shreds of molten rock which surges across the countryside at speeds well in excess of any highway speed limit. When the cloud finally settles, the fragments of molten rock weld themselves together into a solid mass--they are that hot. Such eruptions are explosively violent and every bit as destructive as they sound; everything in the path of those glowing hot clouds of molten rock is instantly incinerated. Several such eruptions have been observed in historic time but none of those produced welded ash beds that went more than a few miles from the vent or covered more than a few square miles. The eruptions that put the welded-ash beds in the John Day Formation must have come from the western Cascades, probably somewhere south of Salem, and the welded-ash ledges extend eastward at least as far as Dayville and cover thousands of square miles. No volcanic eruption on a scale even remotely comparable to that has ever been witnessed in historic time. (page 160)

Most of the Miocene volcanic activity in central and eastern Oregon involved eruption of enormous floods of basalt, some of which covered thousands of square miles with single lava flows having volumes measurable in hundreds of cubic miles. They are well named; they really were floods of basalt lava. No such overwhelming eruptions of basalt have happened anywhere in the world during historic time so we have no eyewitness accounts to help us picture what they were like. (page 161)

80 Smith, R. B., and Christiansen, R. L., 1980, Yellowstone Park as a window on the earth's interior: Scientific American, vol. 242, no. 2, pp. 104-117.

Rocks in and near Yellowstone National Park reveal three phases of explosive Quaternary volcanism. The first, and most catastrophic, ash-flow eruption produced the Huckleberry Ridge Tuff which has a volume of more than 2,500 cubic kilometers. The collapse of the magma chamber created an enormous caldera, now partially covered by more recent deposits, which is believed to extend in an east-west direction 130 kilometers from near Island Park, Idaho, to the central part of Yellowstone. The second phase of catastrophic eruption produced more than 280 cubic kilometers of Mesa Falls Tuff and the 25-kilometer Island Park caldera in eastern Idaho. The final phase expelled 1,000 cubic kilometers of rhyolite pumice and ash producing the Lava Creek Tuff and releasing roof support creating the Yellowstone caldera (75 x 45 kilometers) which occupies about one third of the area of Yellowstone National Park.

81 Ninkovich, D., Sparks, R. S. J., and Ledbetter, M. T., 1978, The exceptional magnitude and intensity of the Toba eruption, Sumatra: an example of the use of deep-sea tephra layers as a geological tool: Bulletin Volcanologique, vol. 41, pp. 286-298.

The Pleistocene eruption of Toba on the island of Sumatra produced what could be the world's largest caldera (100 by 30 kilometers) and volcanic deposits exceeding a volume of 2000 cubic kilometers derived from at least 1000 cubic kilometers of dense rhyolitic magma. Toba is surrounded by an enormous rhyolitic ignimbrite layer covering 20,000 square kilometers, and deep-sea tephra layers from the eruption have been discovered in the Indian Ocean more than 2,500 kilometers from the volcano. The average discharge rate of the volcano during eruption is believed to have been a million cubic meters per second. The eruption column above the

Extrusive and Intrusive Catastrophes

Figure 6

Lava Creek Tuff of Yellowstone National Park, Wyoming. The distinctly stratified air fall volcanic ash (lower half of photo) is overlain by the welded pyroclastic flow deposit (upper half of photo). These explosion products appear to have been ejected in a single volcanic eruption involving more than one thousand cubic kilometers of dense rhyolitic magma which during eruption evicerated an enormous subterrenean chamber into which overlying rock collapsed about three kilometers to form Yellowstone caldera, a 75 by 45 kilometer depression. See Reference 80.

[Courtesy of NOAA/EDS]

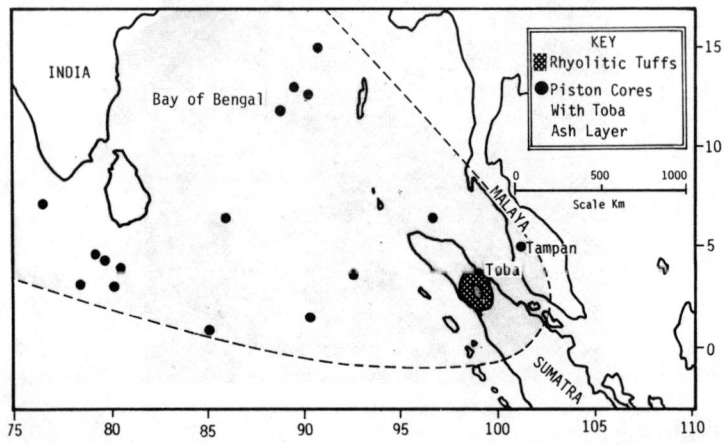

Figure 7

Distribution map of Toba ash layer in the Indian Ocean. At least 1000 cubic kilometers of magma were erupted from Toba on Sumatra producing a discernable seafloor ash layer over 2,500 kilometers from the volcano.
      [Map after Ninkovich, Sparks,
        and Ledbetter, Reference 81]

vent may have exceeded a height of 50 kilometers and stratospheric dust would have affected global climate.

82 Williams, H., and Goles, G., 1968, Volume of the Mazama ash fall and the origin of Crater Lake caldera, in Dole, H. M., ed., Andesite conference guidebook: Oregon Department of Geology and Mineral Industries Bulletin, vol. 62, pp. 37-41.

The ancient explosion of Mount Mazama, the volcanic precursor of Crater Lake caldera, dispersed 30 cubic kilometers of air-fall tephra over an area of almost a million square kilometers.

83 Izett, G. A., and Naeser, C. W., 1976, Age of the Bishop Tuff of eastern California as determined by the fission-track method: Geology, vol. 4, pp. 587-590.

The enormous volcanic eruption, which produced the Pleistocene Bishop Tuff and the Long Valley caldera in California, also deposited air-fall pumice and ash in Nevada, Utah, Wyoming, Colorado, New Mexico, and Nebraska.

84 Barberi, F., Innocenti, F., Lirer, L., Munno, R., Pescatore, T., and Santacroce, R., 1978, The Campanian ignimbrite: a major prehistoric eruption in the Neapolitan area (Italy): Bulletin Volcanologique, vol. 41, pp. 10-31.

A Pleistocene pyroclastic flow deposit around Napoli, Italy is believed to have covered at least 7,000 square kilometers. Ninkovich et al. (1978, reference 81) estimate the dense magma volume of the Campanian eruption at 70 cubic kilometers, so the dispersed volcanic products should exceed 100 cubic kilometers. This prehistoric volcano exceeded the largest historic volcanoes (Taupo, Thera, Tambora, etc.).

85 Mackin, J. H., 1960, Structural significance of Tertiary volcanic rocks in southwestern Utah: American Journal of Science, vol. 258, pp. 81-131.

This paper describes individual pyroclastic flow deposits many hundreds of feet thick having areal extents of as much as 10,000 square miles. The total volume of Tertiary silicic volcanic rocks of the Great Basin of the western United States is reported to be of the order of magnitude of 50,000 cubic miles.

86 Fisher, R. V., 1966, Mechanism of deposition from pyroclastic flows: American Journal of Science, vol. 264, pp. 350-363.

The largest and smallest fragments are believed to be deposited at the base of the flow with little loss of velocity from the main flow. A pyroclastic flow deposit (ignimbrite) from eastern Oregon is described as covering an area in excess of 3,100 square kilometers (1,200 square miles) to an average depth of about 25 meters (80 feet).

Prehistoric Lava Flows

87 Krishnan, M. S., 1968, Geology of India and Burma: Madras, Higginbothams, fifth edition, 536 pp.

The Deccan Traps are Cretaceous and Paleocene basalt strata which form extensive flat-topped, plateau-like features in India representing the most voluminous lava outpourings on earth. The present distribution of the Deccan Traps is about 520,000 square kilometers (200,000 square miles) including Bombay, Kathiawar, Kutch, Madhya Pradesh, Central India and parts of the Deccan. The original extent of the basalt strata is believed by Krishnan to have exceeded 1,500,000 square kilometers (580,000 square miles). The basalt thickness is known at one location to exceed 2,000 meters with individual flows varying from a thickness of a meter to over 36 meters. Some flows have been traced for distances exceeding 100 kilometers. A conservative estimate of Deccan Trap basalt volume using data of Krishnan is 700,000 cubic kilometers (170,000 cubic miles). The extensive lava flows must have affected global climate.

88 Choubey, V. D., 1973, Long-distance correlation of Deccan basalt flows, central India: Geological Society of America Bulletin, vol. 84, pp. 2785-2790.

The Cretaceous Deccan Trap flood basalt flows of India cover an area more than one half million square kilometers with an aggregate volume exceeding one million cubic kilometers. Individual flows have been identified and correlated over distances of 160 kilometers in central India.

89 Swanson, D. A., Wright, T. L., and Helz, R. T., 1975, Linear vent systems and estimated rates of magma production and eruption for the Yakima Basalt on the Columbia Plateau: American Journal of Science, vol. 275, pp. 877-905.

The Rosa Member of the Columbia River Basalt Group (Miocene) of Washington and Oregon contains two extensive basalt strata believed to represent enormous lava flows which "flooded" wide areas in northern Oregon and eastern Washington. The Rosa basalts are estimated to have covered at least 40,000 square kilometers (15,000 square miles) and to have cooled from more than 1,500 cubic kilometers (340 cubic miles) of lava. An enormous fissure system less than 3 kilometers wide and more than 150 kilometers long in southeastern Washington is believed to have been the source of the lava which flowed out at an estimated 100 cubic kilometers per day over the earth distances up to 300 kilometers.

> A variety of evidence suggests that the Roza Member was erupted during a short period of time, perhaps of the order of a few hundred years for the entire member and matter of days for single flows and cooling units. Interflow sediments, except locally derived tephra, have not been found.... (page 883)

The rate of lava eruption is believed to have been up to 100 cubic kilometers per day, about a hundred times greater than the maximum rate of the Laki flow of Iceland (1783), the largest historic lava flow (see Thorarinsson, 1970, reference 68). A Roza lava flow is believed to have released as much as 3×10^{28} ergs of heat energy as it cooled, about a hundred times the yearly global thermal energy released by volcanoes, and must constitute "one of the most significant types of virtually instantaneous energy loss on Earth" (page 900).

90 Wheeler, H. E. and Coombs, H. A., 1968, 100,000 square miles of burning rock: Saturday Review, October 5, pp. 60-63.

This article is a popular account of the technical article (Wheeler and Coombs, 1967, reference 91) presenting evidence for a regionally extensive Pliocene-Pleistocene lava flow in the northwestern United States.

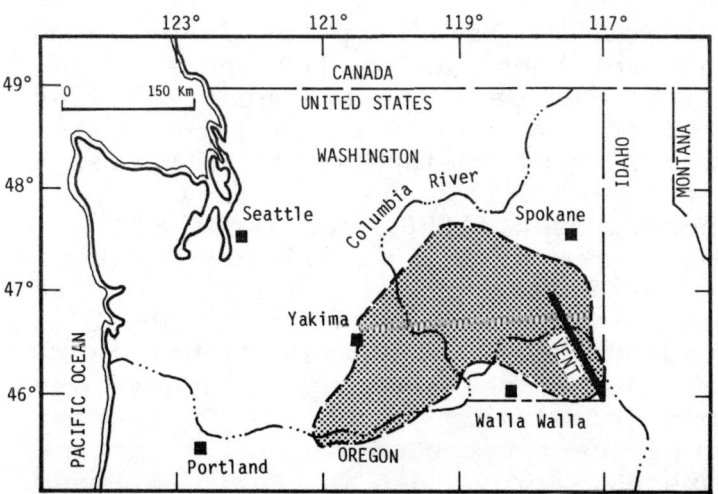

Figure 8

Location map showing the area devastated by the Rosa basalts. A fissure system in eastern Washington flooded a 40,000 square kilometer area with 1,500 cubic kilometers of lava.
[Map after Swanson, Wright and Helz, Reference 89]

Extrusive and Intrusive Catastrophes

> We are now convinced that what we see of the Mesa basalt layer today is only a modest fraction of the original sheet of lava. The major portion of the sheet has been eroded away. The remnants are distributed from near Portland, Oregon, southeasterly to beyond Winnemucca, Nevada, a distance of more than 450 miles. In the northeast-southwesterly direction they occur from near Murphy in the Snake River Valley (south of Boise), Idaho, to near Redding in the northern Sacramento Valley of California. The minimum region encompassed is conservatively estimated at 100,000 square miles. (page 62)

91 Wheeler, H. E. and Coombs, H. A., 1967, Late Cenozoic Mesa Basalt sheet in northwestern United States: Bulletin Volcanologique, vol. 31, pp. 21-43.

This paper is the technical evidence which is considered to document an enormous ancient lava flow involving 350 cubic miles of lava which covered an area of 100,000 square miles. The authors say:

> Numerous occurrences of late Cenozoic (Late Pliocene or early Quaternary) diktytaxitic, olivine basalt in Oregon, northeastern California, northwestern Nevada and southwestern Idaho are interpreted as remnants of a regionally extensive, thin basalt sheet, manifesting a single volcanic event of vast magnitude. (page 21)

Walker and Swanson (1969, reference 92) dispute the idea of a thin, regionally extensive basalt sheet. Even if individual remnants of the postulated lava flow are not continuous, the continuity of basalt observed in the field certainly requires a catastrophic explanation. A popular account is Wheeler and Coombs (1968).

92 Walker, G. W., and Swanson, D. A., 1969, Discussion of paper by H. E. Wheeler and H. A. Coombs, "Late Cenozoic Mesa Basalt sheet in northwestern United States": Bulletin Volcanologique, vol. 32, pp. 581-585.

This paper is a response to Wheeler and Coombs (1967, reference 91) who postulate a thin, regionally extensive lava flow in the northwestern United States. Walker and Swanson dispute the hypothesis by saying that such a flow is "inconceivable" and by noting potassium-argon dates on the basalt ranging from 1.2 to 14.5 million years. Even if the individual remnants of the postulated lava flow are not continuous, the continuity of basalt observed in the field certainly requires a catastrophic explanation.

93 Davis, E. E., 1982, Evidence for extensive basalt flows on the sea floor: Geological Society of America Bulletin, vol. 93, pp. 1023-1029.

Basalt flows exist on the ocean floor and Davis argues that that some may exceed 60 kilometers in horizontal extent.

Hot Intrusive Processes

34 Wyllie, P. J., 1975, The earth's mantle: Scientific American, vol. 232, no. 3, pp. 50-63.

Kimberlite is an unusual intrusive igneous rock containing abundant phenocrysts of olivine and phlogopite in a fine-grained matrix of calcite, olivine and phlogopite. An important accessory mineral in kimberlite is diamond, a mineral form of carbon stable only at very high pressures. The mineral assemblage indicates that kimberlite minerals originated at a depth of 150 to 300 kilometers below the surface within the earth's mantle. A major problem is explaining how kimberlite was intruded into the earth's crust without changing mineral assemblages. Rapid intrusion appears necessary. Wyllie says:

> Evidence indicates that a kimberlite pipe originally rose rapidly through the crust as a fluidized system of solids, molten rock and gases, breaking through to the surface with a tremendous blast in a brief volcanic explosion. (page 55)

35 Anderson, O. L., 1979, The role of fracture dynamics in kimberlite pipe formation, *in* Boyd, F. R., and Meyer, H. O. A., eds., Kimberlites, diatremes, and diamonds: their geology, petrology, and geochemistry: Washington, American Geophysical Union, pp. 344-353.

Diamonds are occasionally found in kimberlites, rocks which are thought to be derived from the earth's mantle (see Wyllie, 1975, reference 94). Diamonds are metastable in magma near the earth's surface and do not crystallize from molten material in the earth's crust. High temperature and pressure at a depth greater than 150 kilometers provide the conditions necessary for diamonds to crystallize. Furthermore, if diamonds are intruded slowly from deep in the earth through the earth's crust, they will pass through the stability field of graphite and will convert to graphite, the stable mineral form of carbon in the earth's crust. The only way to get diamonds to the earth's surface is by rapidly decreasing the temperature and pressure of the diamond-bearing magma by a "quenching" process which bypasses graphitization. This places severe time constraints on the intrusion of diamond-bearing rocks into the earth's crust.

Anderson's paper describes theory and mechanics of catastrophically intruding kimberlite pipes through the earth's crust. He suggests that the material can ascend upward only as fast as a crack opens ahead of it. Low viscosity, carbonate-containing, volatile-rich fluid at the tip of the crack exerts the necessary tension allowing crack formation and fluid injection to accelerate upward at high speed. Anderson believes that diamond-bearing kimberlites may be intruded from a depth of 200 kilometers to the surface in less than 5 hours at an average rate of ascent of 7 meters per second through a crack a meter wide! As the crack and fluid breaks the surface of the earth it would have accelerated to nearly the shear velocity of sound in rock (approximately 4 kilometers per second). The eruption at the surface would be incredible.

96 Mercier, J. C., 1979, Peridotite xenoliths and the dynamics of kimberlite intrusion, in Boyd, F. R., and Meyer, H. O. A., eds., The mantle sample: inclusions in kimberlites and other volcanics: Washington, American Geophysical Union, pp. 197-212.

Laboratory experiments reveal that strained olivine crystals are subject to rapid annealing recrystallization at high temperatures and pressure. The presence of strained olivine crystals in xenoliths from kimberlite indicates magma ascent velocities of 40 to 60 kilometers per hour from the mantle at a depth of 200 kilometers. A large tectonic event must immediately precede kimberlite intrusion in order to strain olivine crystals deep in the earth.

97 Szekely, J., and Reitan, P. H., 1971, Dike filling by magma intrusion and by explosive entrainment of fragments: Journal of Geophysical Research, vol. 76, pp. 2602-2609.

Evidence is presented that igneous dikes have been formed by gas and entrained particles accelerated upward through the earth's crust along fractures at speeds approaching sonic velocities (up to 890 meters per second). Magma is estimated to have traveled upward from a depth of 30 kilometers to the earth's surface in 41 seconds or less.

98 Spera, F. J., 1980, Aspects of magma transport, in Hargraves, R. B., ed., Physics of magmatic processes: Princeton, Princeton University Press, pp. 265-323.

In a section of his paper titled "Magma Ascent Rates: Inferences from Xenoliths," Spera argues that xenoliths (hard rock fragments intruded with the enclosing rock when it was molten) must have been carried upward from great depth at rates which exceeded their settling velocities through the magma. Reasonable assumptions are made for spherical xenoliths in basaltic magmas and settling velocities are usually in the range of tens of centimeters per second. This places a <u>minimum</u> value on the rates of ascent of these magmas and argues for rapid intrusive processes.

19 Irvine, T. N., 1964, Sedimentary structures in layered ultrabasic rocks: American Association of Petroleum Geologists Bulletin, vol. 48, p. 533.

Rapid intrusive processes are indicated by "sedimentary structures" possessed by igneous rocks.

> Many igneous intrusions show layering formed by gravitational accumulation of crystals that is, both in variety and detail, remarkably similar to the bedding of sedimentary rocks. Such layering occurs in most compositional types of intrusions but especially in mafic and ultramafic bodies. The examples considered specifically are from the Duke Island ultramafic complex in the Alaskan panhandle; the rocks are composed of olivine and clinopyroxene but, where layered, look much like graded bedded turbidity current deposits....
>
> Other layering features include: cross-layering, slump and deformation concurrent with accumulation, streamlining of layers over irregularities, load casts, general correlation between layer thickness and particle size, lateral grading, and "diagenetic" recrystallization.
>
> The layering has undoubtedly formed because of magmatic currents during extremely unstable conditions. The currents were probably a convective type of overturn marked by the descent of dense, crystal-laden magma from

the cooling roof and walls of the pluton, and it is likely that they were initiated and perpetuated by repeated slumping. The described phenomena illustrate that features of waterlaid sediments can form in a vastly different environment in terms of the specific gravities of particles and transporting liquid, and the viscosity of the liquid.

100 Gussow, W. C., 1968, Salt diapirism: importance of temperature, and energy source of emplacement, in Braunstein, J., and O'Brien, G. D., eds., Diapirism and diapirs: Tulsa, American Association of Petroleum Geologists, Memoir 8, pp. 16-52.

Salt domes are large mushroom-shaped bodies of salt which have intruded sedimentary strata. The diapiric process which formed them have long baffled geologists. It appears that salt strata at great depth were hot enough for the salt to flow plastically. By a well-known physical principle, lower density salt should penetrate upward through higher density strata. However, as salt penetrates slowly upward, the pressure and temperature in the salt body should decrease and the intrusion process by plastic flow should be arrested. Rapid or catastrophic salt flow would allow the salt body to remain hot and plastic even though it is penetrating a cooler, low pressure region. Gussow says:

> Because there are no recorded observations of the progress of salt diapirism, there is no actual measure of the rate of intrusion. Reports of shale diapirism have been published (Birchwood, 1965), and these indicate that, measured in days, the activity is of rather short duration. Observations of volcanic activity also suggest a rather brief period of intrusion. Balk (1953) correctly estimated the salt temperature at the base of a salt dome at 25,000 ft to be about 570°F (300°C), and he also knew about the remarkable plasticity of salt at high temperatures. However, because he visualized salt-dome growth as a very slow process, he did not realize that 570°F also could have been the temperature of the salt during intrusion and at the time of extrusion. Richter-Bernburg and Schott (1959, p. 88) state, "... the movement is rapid," but do not define "rapid." On the basis of the known physical properties of salt and the fact that salt moved as flows at the surface in Iran, it must be concluded that the salt was hot at the time of extrusion and that extrusion was rather rapid; it would be catastrophic on a geologic time scale. This conclusion is contrary to the published opinions of most geologists. (pages 46 and 47)

Cold Intrusive Processes

01 Roth, A. A., 1977, Clastic dikes: Origins, vol. 4, pp. 53-55.

A clastic dike is a cross-cutting body of sedimentary material which has been intruded into a foreign rock mass. These dikes are most easily recognized where they penetrate horizontal sedimentary strata, but they may occur also in igneous and metamorphic rocks. The process of formation of a clastic dike is analogous to wet sand oozing up between ones toes, but on a much larger scale. In most cases there is clear evidence that the intruded material was not lithified, but soft sediment, and that lithification of the sedimentary source bed of the dike came after intrusion of the dike. Interesting geologic consequences are reported by Roth:

> One series of dikes of special interest to one seeking to determine the age of sediments in the earth is found in the Front Range of Colorado north of Pikes Peak (Cross 1894, Roy 1946, Vitanage 1954, Harms 1965). In this case, sand from the Cambrian Sawatch sandstone has intruded into the Precambrian Pikes Peak granite during the Laramide Orogeny. This orogeny is the main uplift forming the Rocky Mountains which occurred relatively late in geologic time. There is disagreement as to whether the intrusions forming these dikes are from below or from above; in this case the time discrepancy is so great that this point makes little difference. The sandstone dikes contain fragments from the Permian-Pennsylvanian

Fountain Formation, indicating that at least this formation was present at the time of intrusion. On a geologic time scale this represents a period of at least 250 million years during which the Sawatch sandstone remained uncemented. This seems especially unusual since just above the Sawatch are several carbonate layers that could provide an abundant source of cement for the Sawatch. If, as field evidence indicates, intrusion took place during the Laramide Orogeny, the Sawatch sandstone would have had to remain uncemented for more than 400 million years. On the other hand, if, as expected, dikes are formed at approximately the same time as their host rock, or at least the cracking of the host rock during the Laramide Orogeny in the Pikes Peak granite case, then there must not be much time difference between the Cambrian and the Laramide Orogeny which supposedly occurred more than 400 million years later! (pages 53,54)

102 Damberger, H. H., 1970, Clastic dikes and related impurities in Herrin (No. 6) and Springfield (No. 5) coals of the Illinois Basin: in Smith, W. H., et al., Depositional environments in parts of the Carbondale Formation--western and northern Illinois, Illinois State Geological Survey Guidebook Series No. 8, 119 pp.

Clastic dikes have been observed in almost every coal bed in the Pennsylvanian System of Illinois. Under normal compactional processes the

pressure gradient would direct clastic
dikes upward into overlying sediments.
In Illinois coal, however, the dikes
were injected from above. Damberger
believes that earthquake activity
occurred shortly after deposition of
the coal allowing the dikes to be
intruded down into the coal before the
coal was solifidifed. The abundance
of clastic dikes could, therefore,
argue that earthquakes play an important part in compaction of sediments.

03 Pierce, W. G., 1979, Clastic dikes of
Heart Mountain Fault breccia,
northwestern Wyoming, and their
significance: United States
Geological Survey Professional
Paper 1133, pp. 1-25.

Structural features in northwestern Wyoming
indicate that the Heart Mountain fault movement was an extremely rapid, cataclysmic
event that created a large volume of carbonate
fault breccia derived entirely from the lower
part of the upper plate. After fault movement
had ceased, much of the carbonate fault
breccia, here called calcibreccia, lay loose on
the resulting surface of tectonic denudation.
Before this unconsolidated calcibreccia could
be removed by erosion, it was buried beneath a
cover of Tertiary volcanic rocks: the Wapiti
Formation, composed of volcanic breccia,
poorly sorted volcanic breccia mudflows, and
lava flows, and clearly shown in many places by
interlensing and intermixing of the calcibreccia
with basal volcanic rocks. As the weight of

volcanic overburden increased, the unstable water-saturated calcibreccia became mobile and semifluid and was injected upward as dikes into the overlying volcanic rocks and to a lesser extent into rocks of the upper plate. In some places the lowermost part of the volcanic overburden appears to have flowed with the calcibreccia to form dikelike bodies of mixed volcanic rock and calcibreccia. One calcibreccia dike even contains carbonized wood, presumably incorporated into unconsolidated calcibreccia on the surface of tectonic denudation and covered by volcanic rocks before moving upward with the dike....Evidence that the Wapiti Formation almost immediately buried loose, unconsolidated fault breccia that was the source of the dike rock strongly suggests a rapid volcanic deposition over the area in which clastic dikes occur, which is at least 75 km long. (page 1)

Sedimentary Products of Volcanism

104 Bailey, E. H., Irwin, W. P. and Jones, D. L., 1964, Franciscan and related rocks, and their significance in the geology of western California: California Division of Mines and Geology, Bulletin 183, 177 pp.

Chert is a dense and compact rock composed mainly of cryptocrystalline silica. The origin of chert is a subject of controversy among geologists. The theory that bedded chert could form as a marine chemical precipitate from hot volcanic waters lacks a modern counterpart and would appear to require a catastrophic process. The authors favor such a theory for the Franciscan cherts (Jurassic) of California:

> Chert and a distinctive shale occurring with it . . .are believed to be chemical precipitates formed by the reaction of magma and sea water under considerable hydrostatic pressure. They are important as indicators of the oceanic depth in which part of the Franciscan was deposited. Rhythmically interlayered red or green chert and shale form lenses less than 50 feet thick and less than a mile in extent, generally with and above greenstones. . . .
>
> The association of chert-shale lenses with greenstone suggests a genetic relation. The lenses may represent silica, alumina, and iron released by submarine volcanic rocks at the time of volcanic eruption, the eruption occurring at a depth great enough for sea water at

the reactive interface to be heated to a temperature of about 350°C without boiling. At this temperature and at a pressure equal to that of oceanic depths of 13,000 feet, water can dissolve over 1,000 ppm of silica. Such heated, silica-enriched water would rise, be cooled, and quickly become oversaturated with respect to silica. Silica would then be polymerized and precipitated as a gel, apparently along with aluminum and ferrous hydroxide, and it would rain down onto the sea floor forming a mass of impure silica gel. Subsequently, by a process of diffusion and crystallization, layers that superficially resemble normal sedimentary beds would form. Similar though smaller layers were formed experimentally by Davis using sodium silicate and powdered Franciscan shale. This postulated origin for the chert-shale lenses seems to be the only one compatible with all their unusual structural and chemical features, and it implies that deposition of some Franciscan rocks must have been at a depth nearly equivalent to or greater than the average of the Pacific Ocean. (page 6)

The authors have photographs and descriptions of penecontemporaneous chevron folds in thick Franciscan chert sequences which indicate rapid chert deposition and associated tectonism.

105 Lonsdale, P. F., Bischoff, J. L., Burns, V. M., Kastner, M., and Sweeney, R. E., 1980, A high-temperature hydrothermal deposit on the seabed at a Gulf of California spreading center: Earth and Planetary Sciences Letters, vol. 49, pp. 8-19.

A submersible dive on a turbidite-covered spreading axis in Guaymas Basin photographed and sampled extensive terraces and ledges of talc. The rock contains siliceous microfossils, smectite, and euhedral pyrrhotite as well as rather pure iron-rich talc. Sulfur and oxygen isotopes indicate precipitation around a hydrothermal vent, at about 280°C. (page 8)

106 Rode, K. P., 1944, On the submarine volcanic origin of rock-salt deposits: Indian Academy of Sciences, Proceedings, vol. 19, section B, pp. 130-142.

Rode disputes the evaporation theory for the formation of bedded salt deposits and suggests a theory for volcanic origin.

It is quite possible that certain deep synclinal basins occupied by the sea had become seats of such fissure eruptions wherein most of the magmatic gases and vapours got absorbed in the sea water. This accession of magmatic emanations to the salts already existing must have violently disturbed the chemical equilibrium of the sea water, whereas the heat of eruption must have raised the temperature of the sea water abnormally high leading to rapid evaporation and even to boiling. This led to the copious precipitation of particularly those salts which were being added from volcanic vents. (page 138)

107 Sozansky, V. I., 1973, Origin of salt deposits in deep-water basins of Atlantic Ocean: American Association of Petroleum Geologists Bulletin, vol. 57, pp. 589-590.

Sozansky challenges the prevailing opinion that rock salt strata were deposited from brines concentrated by evaporation of seawater. It is nonsense to suppose, says Sozansky, that salt deposits deep on the floor of the Atlantic Ocean formed by complete evaporation of that ocean. Instead, he proposes that salt deposits in deep oceanic areas have accumulated from hot brines originating at great depths in the earth during tectonic movements. His article cites his book and extensive Russian literature supporting his theory. No modern example of salt deposition in marine environments is known to occur by this process. A similar theory for the origin of bedded salt is advocated by Rode (1944, reference 106).

108 Porfir'ev, V. B., 1974, Geology and genesis of salt formations: American Association of Petroleum Geologists Bulletin, vol. 58, pp. 2543-2544.

A Russian geologist reviews the Russian book <u>Geology and Genesis of Salt Formations</u> where V. I. Sozansky challenges the evaporation theory for the origin of rock salt deposits. Instead, Sozansky proposes that salt arises along faults as juvenile hot brines from the earth's mantle. Evidences cited by Sozansky are:

(1) the nearly universal occurrence of salt with extrusive igneous rocks,
(2) the presence of saline solutions from modern juvenile sources in highly faulted areas (Great Rift Valley, Dead Sea, Red Sea, Danakil salt plane, Ol Doinyo Lengai volcano, and Lake Kivu),
(3) the absence of planktonic fossils and traces of other marine organisms in salt versus the large amount of marine organism remains in modern lagoons from which evaporation has left salt,
(4) the absence of magnesium sulfate deposits which form after evaporation in modern lagoons (Gulf of Kara-Bogaz), and
(5) the difficulty in imagining an evaporation basin where the rate of subsidence corresponds exactly with the rate of salt accumulation

to maintain the appropriate restriction to oceanic flooding.

Chapter 5

MASS MOVEMENT CATASTROPHES

Solid materials in the earth's crust move in four ways: by **flowing, sliding, shaking,** and **collapsing.** Many large debris flows and "landslides" move catastrophically by a flow process where large rocks are suspended in a slurry of water, clay, silt and sand. True landslides and rockslides occur by a sliding process where a definite fault surface exists along which displacement occurs. The earth's crust also can be deformed elastically by shaking (earthquake) so that the earth returns almost to its prestressed state with little permanent change (except for folding, faulting, etc. that may result). Finally, the earth can deform by collapse under the direct influence of gravity forming such features as sinkholes and calderas. This chapter reviews the evidence for catastrophic gravity flow deposits (references 109 through 118), reports the research and speculation regarding enormous gravity slide deposits (references 119 to 126), discusses large earthquakes and their effects (references 127 to 134), and describes unusual collapse features (references 135 and 136).

Historic Gravity Flow Deposits

09 Hsü K. J., 1975, Catastrophic debris streams (sturzstroms) generated by rockfalls: Geological Society of America Bulletin, vol. 86, pp. 129-140.

Enormous rockfalls often generate fast-moving streams of debris that have been called "sturzstroms." Hsü describes the characteristics of these debris streams, how they move, and how they deposit material. The speed of a sturzstrom often exceeds a hundred kilometers per hour; the volume is commonly more than a million cubic meters; individual clasts may be as large as a house. The author describes such a phenomenon as "a stream of colliding blocks swimming with terrifying speeds in a sea of small stones and dry rock powder." Movement is by "flow" and is powered by the initial kinetic energy of the fall which produces enough momentum for long distances of transport across flat terrain. Laboratory modeling of these flows shows that the mobility of the "debris stream" is dependent on velocity and volume of moving rock material.

10 Browning, J. M., 1973, Catastrophic rock slide, Mount Huascaran, north-central Peru, May 31, 1970: American Association of Petroleum Geologists Bulletin, vol. 57, pp. 1335-1341.

A catastrophic rock slide involving over 2 million cu m of rock roared down the valley of the Llanganuco River, split into two lobes, one of which killed approximately 2,000 people in the town of Rangrahirca and the other killed approximately 19,000 people in the town of Yungay.

Many thousands of other rock slides within 100 km of Chimbote resulted in many millions of cubic meters of rock forming scree slopes at the foot of steep hills and mountains.

The slide was triggered by an earthquake 85 km away, at a depth of 54 km. In the upper and steepest drop the slide fell and swept as a nearly friction-free unit mass a distance of 14.5 km, a vertical drop of 13,700 ft., in 3 minutes. Over the next 50 km the slide moved as a fluid mass, covering the distance in 2 hours. Deposition from the slide was velocity-dependent. (page 1335)

111 Kojan, E., and Hutchinson, J. N., 1978, Mayunmarca rockslide and debris flow, Peru, in Voight, B., ed., Rockslides and avalanches: New York, Elsevier, pp. 315-361.

This gigantic rockslide and debris flow occurred in the Andes of Peru in April 1974 and had a volume of about one billion cubic meters (one cubic kilometer). The slide mass made a dam 3.8 kilometers long and 150 meters high across a valley producing a lake 38 kilometers long, which, 44 days after the rock-slide, was catastrophically breached causing a flood. The flood wave was 20 meters high 100 kilometers downstream from the slide dam. The initial speed of the flood water is estimated at between 15 and 30 kilometers per hour. Maximum discharge through the breached dam was nearly 10,000 cubic meters per second.

Prehistoric Gravity Flow Deposits

112 Chadwick, A. V., 1978, Megabreccias: evidence for catastrophism: Origins, vol. 5, pp. 39-46.

Chadwick defines <u>megabreccia</u> as any sedimentary deposit in which angular fragments of rock in excess of one meter in diameter occur as conspicuous components. Different mechanisms for rock fragment transport are outlined and examples of very large movements requiring very great energy are documented. A megabreccia flow in the Cambrian Tapeats Sandstone of the Grand Canyon contains enormous quartzite boulders. As a survey of mass movement processes, this paper should challenge researchers to formulate models which explain the field evidence.

> The presence of various kinds of megabreccias in the geologic column, showing in some cases the transport of extremely large clasts, indicates energy levels on a scale that staggers our imagination. Their common occurrence in major portions of the geologic column of some localities indicates significant catastrophic activity in the past not readily explanable in terms of contemporary processes. (page 44)

Figure 9

Megabreccia stratum at the base of the Tapeats Sandstone in the bottom of the Grand Canyon. The large boulder above and to the right of the man in the lower left corner has a diameter of five meters and weighs nearly two hundred tons. The entire boulder bearing stratum, including the two hundred ton boulder, is believed to have been deposited from a catastrophic underwater mass flow.

[Photo by S. A. Austin]

13 Johns, D. R., Mutti, E. Rosell, J., and Seguret, M., 1981, Origin of thick, redeposited carbonate bed in Eocene turbidites of the Hecho Group, south-central Pyrenees, Spain: Geology, vol. 9, pp. 161-164.

A limestone bed of the Eocene Hecho Group of northern Spain is interpreted as a single-event, sediment gravity flow deposit initiated by an earthquake. The bed is called the "Roncal Unit," commonly exceeds 100 meters thickness, and can be followed along outcrop for 75 km. The volume of the gravity flow deposit is estimated conservatively to be 60 cubic kilometers.

14 Bailey, E. B., and Weir, J., 1932, Submarine faulting in Kimmeridgian times: east Sutherland: Transactions of the Royal Society of Edinburgh, vol. 57, pp. 429-454.

This classic work by Sir Edward Bailey describes beds of boulders contained in a marine black shale sequence from the Jurassic of Scotland. Some of the boulders are more than 100 feet long. The boulders are thought to have been produced on a submarine fault scarp and distributed by earthquake over the ocean floor. The resulting tsunami, which would certainly have been generated by the earthquake and submarine faulting, deposited shallow marine organisms and sediment on the boulders. Black shale is shown to intertongue with beds of boulders. Penecontemporaneous deformation structures are common. Clastic dikes were injected into the sedimentary deposits by earthquake action.

115 Conaghan, P. J., Mountjoy, E. W., Edgecombe, D. R., Talent, J. A., and Owen, D. E., 1976, Nubrigyn algal reef (Devonian), eastern Australia: allochthonous blocks and megabreccias: Geological Society of America Bulletin, vol. 87, pp. 515-530.

Devonian megabreccia deposits of New South Wales, Australia, contain limestone clasts as much as one kilometer across transported by enormous subaqueous debris flows. These deposits had been interpreted as in situ "algal reefs" or "bioherms" by earlier investigators.

116 Lindsay, J. F., 1966, Carboniferous subaqueous mass-movement in the Manning-Macleay Basin, Kempsey, New South Wales: Journal of Sedimentary Petrology, vol. 36, pp. 719-732.

Fifty-five billion cubic meters of coarse sedimentary rock of New South

Figure 10

Mass flow deposits of Lycium Wash., San Diego County, California. Turbidite sandstone (lower half of photo) is overlain by megabreccia (upper half of photo) containing large shale rip-up clasts (upper left corner of photo) and larger granite boulders (upper right corner of photo). The boulders and rip up clasts were moved as a flow suspended in a dense slurry of water and sand, then were deposited.
[Photo by S. A. Austin]

Wales, Australia, originally interpreted as deposited by Gondwanan glaciers, are believed, instead, to have been accumulated by subaqueous mudflows.

117 Schermerhorn, L. J. G., 1974, Late Precambrian mixtites: glacial and/or nonglacial?: American Journal of Science, vol. 274, pp. 673-824.

Late Precambrian boulder and gravel deposits have been interpreted as "tillites" deposited by continental glaciers according to some geologists. Schermerhorn disputes this conclusion calling the deposits "mixtites" attributing most to submarine mass flows adjacent to active tectonic areas.

118 Lucchitta, B. K., 1978, A large landslide on Mars: Geological Society of America Bulletin, vol. 89, pp. 1601-1609.

Photography from the Viking mission reveals a number of spectacular landslide deposits adjacent to walls of large chasmas on Mars. The enormous landslide deposit on the south wall of Gangis Chasma contains at least 100 billion cubic meters (100 cubic kilometers) of material, involved a maximum vertical drop of 2,000 meters, and extended 60 kilometers away from the source area. The speed of more than 100 kilometers per hour is estimated, and the total energy is believed to be 1.5×10^{25} ergs making it the most energetic mass movement deposit yet documented.

Historic Gravity Slide Deposits

119 Voight, B., 1973, The mechanics of retrogressive block-gliding, with emphasis on the evolution of the Turnagain Heights Landslide, Anchorage, Alaska, in DeJong, K. A., and Scholten, R., eds., Gravity and tectonics: New York, John Wiley, pp. 97-121.

The Alaskan earthquake of March 27, 1964, caused severe damage to the city of Anchorage including its residential development Turnagain Heights. A thin clay layer buried at a depth of 100 to 150 feet liquified during the earthquake producing an essentially frictionless surface allowing approximately 130 acres to become involved in a complex horizontal slide on a slope of 2 degrees. The horizontal displacement was as great as 2000 feet and must have occurred in less than 9 minutes. The slide destroyed 75 homes, killed 3 people, and involved about 12.5 million cubic yards of earth. The slide mechanism may provide insight into the much larger Heart Mountain Rockslide (Voight, 1974, reference 123).

150 Catastrophes in Earth History

Figure 11

Aerial view of Turnagain Heights landslide of Anchorage, Alaska. Catastrophic ground rupture and horizontal sliding, which destroyed a subdivision, was caused by the earthquake of March 27, 1964. See Reference 119.
[Courtesy of NOAA/EDS]

20 Bombolakis, E. G., 1981, Analysis of a horizontal catastrophic landslide, *in* Carter, N. L., Friedman, M., Logan, J. M., and Stearns, D. W., eds., Mechanical behavior of crustal rocks: Washington, American Geophysical Union, monograph 24, pp. 251-258.

The Hartford Dike slide, a small, catastrophic horizontal landslide, did not move along a zone of liquified material (as the Turnagain Heights landslide) but appears to have slid along a surface of high fluid pressure which could not be dissipated during the brief event. This historic landslide is useful in suggesting how certain overthrust faults, decollements, and "gravity tectonic" slides could occur. The author says, "No mechanism yet has been documented as to how the shear resistance along glide surfaces is reduced sufficiently to account for the displacements in overthrusts, decollements, and low-angle 'gravity tectonic' slides." (page 251)

Prehistoric Gravity Slide Deposits

121 Tazieff, H., 1976, Horizontal landslides during the 1960 Chile earthquake: Catastrophist Geology, vol. 1, no. 2, pp. 27-32.

A French geologist describes catastrophic horizontal landslides during the 1960 Chile earthquake and extrapolates the effects to the prehistoric processes which formed large alpine nappes.

> To sum up, we have learned that during the short interval in which a great quake is active, dry ground can break off along a fracture and move at relatively high speeds over considerable horizontal distances. Shearing caused the fracturing of layers into separate blocks, several thousands of cubic metres each; other layers were reduced to small fragments that were transported chaotically....
>
> In the face of such overwhelming evidence for horizontal displacements, we are entitled--within reasonable limits--to make several extrapolations. The conditions prevailing in this piedmont region of the Cordillera were not at all favourable to great mass movements: the distance from the epicentre was more than 200 km, the ground was dry, and the surfaces over which the avalanches moved were horizontal. In spite of this, considerable masses were transported to places up to 1 km from their point of origin. So we may wonder what might happen to wet material, either on the continents or on the ocean floor. Over how many kilometers thick neritic or bathyal sediments can be transported when they are in or near the epicentral zone?....

An even greater mystery are the alpine nappes, these folds so exaggerately overturned and so stretched out that their lithified layers have become transported scores of kilometers from their original location. Next to the conventional hypotheses of compression and of gravity flow (écoulement par gravité), we are led to consider henceforeward the possibility that they were formed during world quakes. I think this is the mechanism responsible at least for those nappes without roots--piles of layers whose provenance can no longer be identified. So far, their genesis has remained a mystery.

The importance of earthquakes as an agency in geomorphology, sedimentology and tectonics now seems established beyond reasonable doubt, and it would be interesting to examine a host of unexplained features of land forms, sediments and fold-mountains in the light of the lessons taught by the Chile earthquake. (pages 31 and 32)

122 Lemoine, M., 1973, About gravity gliding tectonics in the Western Alps, in DeJong, K. A., and Scholten, R., eds., Gravity and tectonics: New York, John Wiley, pp. 201-216.

A nappe is a large sheet-like mass of rock which has been displaced horizontally from its original position by tectonic forces. Recumbant folding and overthrust faulting are the primary mechanisms believed to move nappes. The Western Alps are believed to contain many nappes, some of which exceed 800 cubic kilometers volume, which are claimed to have been displaced as much as a hundred kilometers by the force of gravity along nearly horizontal fault surfaces. In a section of his paper titled "Velocity, Kinetic Energy, Obstacles, and the Principle of Uniformitarianism," Lemoine presents the mechanical paradox of large-scale nappe displacement on overthrust faults and admits that catastrophic sliding of nappes needs to be considered.

> Let us remember that the catastrophic Vaiont slide of 1963 has been proved to have had a maximum velocity of 60 km/hr (Selli and others, 1964). However, we are not sure that such a catastrophic event, involving a small volume of rocks (0.3 km^3) and a somewhat steeper slope (45° to 5-10°), may be compared to the gliding of large cover nappes (800 km^3 for the Breche nappe of Chablais only!)....
>
> But the Vaiont slide has been observed before, during, and after its motion, whereas true large gliding nappes in motion are unknown to date. If nappe gliding is a very slow phenomenon, it is possible that nappes are still gliding somewhere on the earth, even if we can neither measure nor observe their movement....
>
> If the velocity is really low, the kinetic energy of the nappe will be very small. For example, the kinetic energy of the whole assemblage of Prealpine nappes of the Chablais and the Préalpes Romandes (6600 km^3), moving at a rate of 1 m/yr, is 1000 times less than the

kinetic energy of a single automobile moving at a speed of 100 km/hr. Would, in that case, large obstacles such as hills be pushed and dragged along? A depression, bottoming out to a lesser slope, will perhaps stop the nappe (Fig. 11), and a simple calculation shows that this will occur after only a very short time (a fraction of a second). The very slowly gliding geological body will certainly not continue by its own momentum.

In conclusion, we must confess that we cannot say much about this problem of velocity and kinetic energy, except that it cannot be neglected. Perhaps the tectonic slides were catastrophic, but in that case uniformitarianism does not hold, for we do not observe them at present, at least not true large ones. If, on the contrary, the nappes moved slowly, erosion has time to act and irregularities of the slope may perhaps stop the movement. (pages 212, 213, copyright 1973 by John Wiley & Sons, Inc.)

123 Voight, B., 1974, Architecture and mechanics of the Heart Mountain and South Fork rockslides, in Voight, B., ed., Rock mechanics: the American Northwest: Pennsylvania State University, College of Earth and Mineral Sciences, pp. 26-36.

The Heart Mountain area of northwestern Wyoming contains at least four dozen structurally distinct plates which are scattered over an area of about 2000 square kilometers (770 square miles). These plates lie on fault breccia and contain sheared igneous dikes (which do not penetrate through the fault) and other evidence of massive horizontal displacement. Voight believes that the displaced plates were originally assembled into a single mass a kilometer thick which covered about 1300 square kilometers (500 square miles). The plates are believed to have slid at a velocity of about 185 kilometers per hour several tens of kilometers toward the southeast on a slope of about one degree! Evidence of rapid slide comes from the need to move the plates great distances on such low slope, from the properties of the fault surface, and from the similarity to the Turnagain Heights Landslide of Alaska (see Voight, 1973, reference 119). Prostka (1978, reference 124) reviews theories of fault mechanisms. Pierce (1979, reference 103) describes the fault breccia.

24 Prostka, H. J., 1978, Heart Mountain Fault and Absaroka volcanism, Wyoming and Montana, U.S.A., In Voight, B., ed., Rockslides and avalanches: New York, Elsevier, pp. 423-437.

The Heart Mountain area of northwestern Wyoming contains large, overthrust faulted rock bodies believed to be emplaced by a catastrophic rockslide (see Voight, 1974, reference 123). Prostka believes that basal friction was significantly reduced by volcanic gas injected catastrophically into the fault. Also considered are the notions of catastrophic gravitational sliding of the rock body after being launched upward by earthquake and the notion of reduction of friction due to build up of fluid pressure in the rocks.

125 Moore, D. G., Curray, J. R., and Emmel, F. S., 1976, Large submarine slide (olistostrome) associated with Sunda Arc subduction zone, northeast Indian Ocean: Marine Geology, vol. 21, pp. 211-226.

A seismic-refraction-profile survey of the northeastern Indian Ocean floor has located a large submarine slide complex at the base of the continental slope off the Bassein River in the Bay of Bengal southwest of Burma. The complex consists of a large mudflow deposit overlain by coherent slide blocks of stratified sediment. Individual slide blocks are up to 360 meters thick with a length of up to 2.8 kilometers which have been transported distances up to 55 kilometers. The total area of the slide and mudflow is 4,000 square kilometers and the total volume is over 900 cubic kilometers. The investigators believe the slide complex moved catastrophically when it received impetus from a large earthquake.

126 Moore, J. G., 1964, Giant submarine landslides on the Hawaiian Ridge: United States Geological Survey Professional Paper 501-D, pp. D95-D98.

Two large submarine landslides exist on the northeastern slope of the Hawaiian Ridge northeast of Oahu, one occupying an area of 8,000 square kilometers and the other an area of 4,000 square kilometers. Both slides are marked by a concave escarpment at their upper ends and contain flat-topped, tilted seamounts up to 25 kilometers long. Other evidence is described that indicates that faulting on the west flank of Mauna Loa and south flank of Kilauea are caused by similar landsliding on a huge scale.

Historic Earthquakes
and their Effects

127 Broadhead, G. C., 1902, The New Madrid
Earthquake: American Geologist,
vol. 30, pp. 76-87.

The Mississippi River Valley was violently shaken in 1811 and 1812 by earthquakes centered near New Madrid, Missouri. The intensity of the tremors exceeded the historic California earthquakes and were felt over an area of 2 million square miles and as far as 1,000 miles away in Quebec. Earthquake waves were seen on the ground surface and aftershocks continued for two years. In addition to the violent earth shaking were elevation and depression of large tracts of land, opening of huge fissures, spouting of solids from cracks, explosions and flashes of ejected gases, seiching of the Mississippi River, and creation of large lakes. Broadhead's account of the earthquakes includes early reports, some by eyewitnesses. Eliza Bryan, a resident of New Madrid, described the events of February 7, 1812:

> The awful darkness of the atmosphere which now, as formerly, was saturated with sulphurous vapor, and the violence of the tempestuous thundering noise that accompanied it, together with the other phenomena mentioned as attendant on the former ones, formed a scene, the description of which would require the most sublime and fanciful imagination. At first the Mississippi seemed to recede from its banks, its waters gathered up like mountains,

leaving boats high upon the sands. The waters then moved inward with a front wall 15 to 20 feet perpendicular and tore the boats from their moorings and carried them up a creek closely packed for a quarter of a mile. The river fell as rapidly as it had risen, and receded within it banks with such violence that it took with it the grove of cottonwood trees which hedged its borders. (page 77)

Fissures, ejected debris and topographic changes are described by Eliza Bryan:

During all the hard shocks the earth seemed horribly torn to pieces. The surface of hundreds of acres was from time to time covered over for various depths by sand which issued from the numerous fissures. Some of these fissures closed up immediately after they had vomited forth sand and water. What seemed to be coal was thrown up with the sand in some places. It was impossible to say how deep the fissures were. The site of the town evidently settled down fifteen feet, but a half mile below the town there does not seem to be any alteration in the river bank, but a little way back, the numerous large ponds which covered a large part of the country were nearly dried up. (pages 77 and 78)

Professor J. W. Foster relates the geologic effects:

Fissures would be formed 600 to 700 feet long and 20 to 30 feet wide through which water and sand spouted 40 feet high. There issued no burning flames but flashes such as would result from an explosion of gas, or from passing of electricity from cloud to cloud. Oak trees would be split in the center and for 40 feet up the trunk, one part standing on one side of a fissure, the other part on the other .... Near the St. Francis river there is a great deal of sunk land, caused by the earthquake of 1811. Here are large trees submerged ten or twenty feet beneath the water. Previous to the earthquake keel boats would come up the St. Francis river and pass into the Mississippi three miles below New Madrid. The bayou is now dry land. From one of the fissures there was ejected the cranium of an extinct species of ox. In Reelfoot lake the fisherman floats his canoe above the branching submerged tops of cypress trees. Reelfoot lake in Obion Co., Tennessee, nearly 20 miles long and seven broad owes its origin to the sinking of the ground during this period. (pages 78 and 79)

128 Plafker, G., 1965, Tectonic deformation associated with the 1964 Alaska earthquake: Science, vol. 148, pp. 1675-1687.

The earthquake of March 27, 1964, produced observable crustal deformation, or probable deformation, over an area more than 170,000 square kilometers, making its areal effect more extensive than any other single earthquake experienced by geologists.

129 Foster, H. L., and Karlstrom, T. N. V., 1967, Ground breakage and associated effects in the Cook Inlet area, Alaska, resulting from the March 27, 1964, earthquake: United States Geological Survey Professional Paper 543-F, pp. F1-F28.

The 1964 Alaskan earthquake caused ground failure, the formation of cracks, and the injection of clastic dikes. The largest cracks were as much as 30 feet across and 25 feet deep.

130 Kanamori, H., 1977, The energy release of great earthquakes: Journal of Geophysical Research, vol. 82, pp. 2981-2987.

The earthquake magnitude-energy equation devised by Gutenberg and Richter is

$$\log E = 1.5 M_s + 11.8$$

where E is the wave energy of the earthquake in ergs and M_s is the conventional magnitude of the 20-second period surface wave from the seismogram. This equation works well in estimating the energy of large earthquakes, but because of the saturation of the ordinary magnitude scale, underestimates the wave energy of the great earthquakes. The "saturation" results from damping of increases in amplitude where the fault rupture length exceeds about 37 miles, approximately the wave length of the 20-second period waves. A better magnitude measure for the great earthquakes is M_o, the "seismic moment," which is defined by UDS where U is the rigidity, D is the average offset on the fault, and S is the surface area of the fault. Study of large and great earthquakes shows a linear relationship between log M_o and log S indicating the equation

$$M_o = 1.23 \times 10^{22} S^{1.5} \text{ dyne centimeter}$$

where S is in square kilometers which can be easily estimated from the aftershock area. Kanamori gives the best estimate of the energy of a great earthquake as

$$E = M_o/(2 \times 10^4)$$

if a complete stress drop is assumed to occur in the earthquake. The energies of some great earthquakes are given in the following table.

Earthquake	M_s	M_o, 10^{27} dyn cm	E (in ergs)
Chile, May 1960	8.3	2000	1.0×10^{26}
Alaska, Mar. 1964	8.4	820	4.1×10^{25}
Aleutian Islands, Mar. 1957	8 1/4	585	2.9×10^{25}
Kamchatka, Nov. 1952	8 1/4	350	1.8×10^{25}
Ecuador, Jan. 1906	8.6	204	1.0×10^{25}

The 1960 Chile earthquake, possibly the largest in the last 200 years, had an estimated energy of 1.0×10^{26} ergs, but should have been exceeded by earthquakes generated by large asteroid or comet impacts (see Clube and Napier, 1982, reference 134.)

131 Chinnery, M. A., and North, R. G., 1975, The frequency of very large earthquakes: Science, vol. 190, pp. 1197, 1198.

A theoretical maximum earthquake magnitude is believed to exist for earthquakes generated by relief of strain in the earth's crust. The authors argue that the largest, most catastrophic earthquakes have not yet been recorded on seismographs. An earthquake with seismic moment $M_o = 10^{31}$ dyne-centimeters (an energy of about 5×10^{26} ergs) should have a mean return period of 50 years, but such an earthquake is about four times larger than the Chile earthquake of 1960, the largest recorded on seismograph records.

Prehistoric Earthquakes
and their Effects

132 Robinson, J. W., and Threet, R. L., 1974, Geology of the Split Mountain area, Anza-Borrego Desert State Park, eastern San Diego County, California, in Hart, M. W., and Dowlen, R. J., eds., Recent geological and hydrologic studies eastern San Diego County and adjacent areas: San Diego, San Diego Association of Geologists, pp. 47-56.

The south side of Vallecito and Fish Creek Mountains east of San Diego, California is flanked by a southward dipping pediment showing that the original drainage of Fish Creek was toward the south. The present drainage, however, is toward the north through the mountains! Robinson and Threet propose that the drainage of Fish Creek was captured along a north-trending fault that split a thousand-foot-high mountain during a single enormous earthquake.

133 Schultz, P. H., and Gault, D. E., 1975, Seismic effects from major basin formations on the Moon and Mercury: The Moon, vol. 12, pp. 159-177.

Enormous seismic events generated by asteroid impacts created grooved and hilly terrains on the Moon and Mercury. Wide grooves near the Mare Ingenii region of the Moon (160°E, -34°)

are antipodal to the Imbrium Basin (20°W, +35°), a 10^{34} erg asteroid impact structure. Similar disrupted terrains antipodal to the Orientale Basin (the Moon) and the 1300-kilometer-diameter Caloris Basin (Mercury) formed from P-waves and surface waves which converged and were amplified at the antipode producing surface displacements of many meters. Each of these seismic events would have been a million times as energetic as the largest historic earthquakes recorded on seismographs.

34 Clube, S. V. M., and Napier, W. M., 1982, The role of episodic bombardment in geophysics: Earth and Planetary Science Letters, vol. 57, pp. 251-262.

The impact of a large asteroid or comet having a mass of 1×10^{17} grams would release approximately 1×10^{30} ergs (approximately equivalent to 2.5×10^7 megatons of TNT) and produce a crater the size of Sudbury or Popigai (approximately 100 kilometers diameter). Clube and Napier describe the earthquake that would accompany such a large impact.

> About 10^{-2} of the impact energy of a missile will manifest itself as seismic waves, which for a 10^{17}-g missile yields 2.5×10^5 Mt of seismic energy as against 100-500 Mt for a major earthquake: This will propagate globally and it

is therefore very likely that weak crustal structures will be violently disturbed, initiating volcanic and secondary earthquake activity on a worldwide scale. (page 257)

Collapse Features

135 LaMoreaux, P. E., and Warren, W. M., 1973, Sinkhole: Geotimes, vol. 18, no. 3, p. 15.

An enormous sinkhole in limestone formed suddenly on December 2, 1972, in Shelby County, Alabama. The collapse structure is 140 meters long, 115 meters wide, and 50 meters deep occurring within a 16 square kilometer area with about 1,000 other sinkholes. This sinkhole is one of the largest in the United States.

136 Lipman, P. W., 1976, Caldera-collapse breccias in the western San Juan Mountains, Colorado: Geological Society of America Bulletin, vol. 87, pp. 1397-1410.

Four large Oligocene calderas in the western San Juan Mountains of Colorado contain megabreccias formed by the collapse and slumping of the original volcanoes. Individual clasts are as much as 500 meters in length.

Chapter 6

WATER CATASTROPHES

Floods are the world's most widespread, prevalent, and damaging type of natural disaster. In addition to the severe consequences of high rainfall, water can cause significant erosion or deposition from disturbances initiated by impulsive events such as storms, earthquakes, landslides, volcanic explosions, meteorite impacts, and breachings of natural dams. Geologists have not adequately explored the effects of water catastrophes in the geologic record, but have largely ignored them. This chapter describes various types of historic water catastrophes and their geologic effects (references 137 through 155), searches for evidences of rapid deposition of sedimentary rocks (references 156 through 173), explores a variety of ancient geomorphic features as products of water catstrophes (references 174 through 184), and explains the physics of asteroid impact with water (references 185 through 188).

Historic Water Catastrophes

37 Ball, M. M., Shinn, E. A., and Stockman, K. W., 1967, The geologic effects of Hurricane Donna in south Florida: Journal of Geology, vol. 75, pp. 583-597.

This paper describes the massive erosional and depositional effects of one of the most violent hurricanes of the century (Donna, September 1960). The storm tide was up to 3.7 meters (12 feet), which, during a period of less than 6 hours, flooded an area up to 8 kilometers (5 miles) inland along the southern end of the Florida mainland. As the tide ebbed, a layer of lime mud up to 15 centimeters (6 inches) thick was deposited on the mainland and on islands in Florida Bay. Cores and photographs taken before and after the hurricane allowed the authors to compare the catastrophic storm effects with the normal day-to-day processes. The amount of boulder-size rubble formed by the hurricane surf far exceeded the amount produced by day-to-day processes between hurricanes. Large amounts of carbonate, skeletal sand were transported. The carbonate sand storm deposits had cross-bedding resembling that produced by normal tides and surf currents except that the storm deposits were cross-bedded on a larger scale. The layered lime mud desposited by the ebb of the storm tide was common on supratidal flats (above the normal high tide line). Mounds of muddy sediment were not eroded by the storm waves or currents.

138 Hayes, M. O., 1967, Hurricanes as geological agents: case studies of Hurricanes Carla, 1961, and Cindy, 1963: University of Texas, Bureau of Economic Geology, Report of Investigation No. 61, 56 pp.

The importance of rare and catastrophic events is revealed by Hurricane Carla, 1961, which was able to reshape the Texas coastline which had changed only slightly in years before the storm.

> As the storm moved landward, it picked up mollusc shells, rock fragments, coral blocks, and other materials from depths as great as 50 to 80 feet and deposited them on the barrier island. After the storm passed inland, strong currents spilled out of the numerous hurricane channels out into the island by the storm-surge flood. These currents deposited a thin layer (0.5 to 1.5 inches) of sand over what was previously sandy mud bottom out to depths of 60 feet and a graded layer of fine sand, silt, and clay (<u>a turbidite</u>) farther out on the shelf. The storm removed a belt of foredunes 20 to 50 yards wide from the seaward side of Padre Island and left the foredune ridge with wave-cut cliffs up to 10 feet high. The formation of a broad, flat <u>hurricane beach</u> drastically altered the beach profile. The landward side of the barrier island (wind-tidal flats) received much washover material containing surf zone

and beach molluscs. The storm also submerged high-level mud flats along the landward side of Laguna Madre and covered them with a fresh layer of mud. (page 1)

Observations of the effects of storms have application for interpreting ancient sedimentary deposits.

> Some important stratigraphic implications of these observations include: (a) hurricanes can mix environment-sensitive faunas from a variety of environments into a single sedimentary deposit; (b) hurricanes can play a primary role in sediment transport in nearshore environments; (c) hurricanes displace sedimentary processes such that sediment textures and structures normally related to a particular process (e.g., fluvial channel flow) may occur in alien areas; and (d) a great deal of energy is expended sporadically in nearshore environments rather than in a uniform, constant manner. (page 1)

Both the insignificant and the extraordinary are the architects of the sedimentary record. Interpreting the record depends on how the scientist visualizes these extraordinary events.

> The reconstructive ability of the paleogeographer and paleoecologist depends on how well he can conjure up a mental image of an environment and its associated sedimentary processes and apply this image to an outcrop of ancient sediments. The introduction of alien sedimentary processes into an area by catastrophic

storms taxes the imagination even more, because these infrequent visitations of exceptionally effective processes must be pictured against a background of all the normal processes that are at work in the area. (page 52)

139 Doehring, D. O., and Vierbuchen, R. C., 1971, Cave development during a catastrophic storm in the Great Valley of Virginia: Science, vol. 174, pp. 1327-1329.

One of the most violent hurricanes of the century dropped 25 to 28 inches of rainfall at several locations in Virginia during an 8 hour period. Subterranean drainage through Cave Springs Cave near Lexington, Virginia, created flow velocities in excess of 10 feet per second and caused significant erosion and corrosion. Effects produced within the cave include solution of scallops high on the cave walls, erosion of potholes above the normal water level, and deposition of terrace-like deposits of conglomerate on the sides of the cave. Joints feeding the cave were enlarged and an enormous load of sediment must have been transported through the cave. The authors believe that one catastrophic storm can perform more work within a cave than many centuries of more typical erosion.

Cory, H. T., 1913, Irrigation and river control in the Colorado River delta: Transactions American Society of Civil Engineers, vol. 76, pp. 1204-1453.

This detailed account describes one of the world's largest man-made disasters--the unrestrained flooding by the Colorado River in 1905-1907 of the Imperial Valley and the formation of the Salton Sea of California. In February 1905, during a flood on the Colorado River, irrigators were unable to control water entering a diversion canal. Soon the entire river was diverted. The water split and formed two channels (the New and Alamo Rivers) which eroded extensive gorges. By the time the breach was closed in February, 1907, a large lake, the Salton Sea, and an extensive delta had formed. In March, 1907, when the lake reached its maximum height, it was 45 miles long and up to 20 miles wide, covering an area of about 520 square miles. The geologic effects are described by Cory:

> The effect of this flood, in a geological way, was of extraordinary interest and very spectacular. In 9 months the runaway waters of the Colorado had eroded from the New and Alamo River channels and carried down into the Salton Sea a yardage almost four times as great as that of the entire Panama Canal. The combined length of the channels cut out was almost 43 miles, the average width being 1,000 ft., and

the depth 50 ft. To this total of from 400,000,000 to 450,000,000 cu. yd. must be added almost 10 % more for side canyons, surface land erosion, etc. Very rarely, if ever before, has it been possible to see a geological agency effect in a few months a change which usually requires centuries. (page 1324)

141 Kiersch, G. A., 1964, Vaiont reservoir disaster: Civil Engineering, vol. 34, no. 3, pp. 32-39.

The world's worst dam disaster occurred on October 9, 1963, at Vaiont, Italy where more than 2,000 people were killed. About 240 million cubic meters of rock and soil slid catastrophically into the reservoir creating an enormous water wave which ran up 260 meters (850 feet) above the reservoir level on the opposite side. Another wave overtopped the crest of the dam by a height of some 100 meters (328 feet) inflicting complete destruction downstream within a span of 7 minutes. An enormous compressive air blast initiated by the slide preceded the water wave and flooding. The double-curved, thin-arch dam was stressed far in excess of its design but remained structurally sound.

42 Coleman, P. J., 1978, Tsunami sedimentation, in Fairbridge, R. W., and Bourgeois J., The encyclopedia of sedimentology: Encyclopedia of Earth Science Series, vol. VI, Stroudsburg, Pennsylvania; Dowden, Hutchison & Ross, Inc., 901 pp.

Coleman describes the tremendous erosional and sedimentary effects that have been observed and which can be reasonably postulated to be produced by a tsunami ("tidal wave"). In shallow ocean areas invading waves can exceed tens of meters high, extend inland several kilometers, initiate water currents exceeding 10 knots persisting for hours, and cause destructive sloshing and seiche movements in partly enclosed bays and harbors. Tsunami-induced changes observed in ocean bottom depths indicate erosion and deposition of sediments in thicknesses of 20 meters or more over areas as much as a square kilometer. Probable results of the tsunami process discussed by Coleman include:

1. deposition of "wildflysch," "tilloidal," and "chaotic" sedimentary sequences,
2. formation of sediments possessing current indicators conflicting strongly with the direction of slope,

3. production of turbiditic features including graded bedding, flute casts, and "pelagic" layers,
4. interlayering of terrestrial and marine organisms,
5. deposition of paraconglomerates and edgewise conglomerates,
6. reworking of deltaic sediments,
7. breaking and dumping of large amounts of reef material on the seaward and shoreward sides of reefs,
8. erosion of submarine canyons,
9. generation of turbidity currents.

143 Stewart, C., 1883, Letter: Knowledge, vol. 4, p. 395.

A seiche is a free or standing-wave oscillation on the water surface of an enclosed or semi-enclosed basin such as a lake, landlocked sea, bay or harbor. The change in water height may exceed a few meters and can have periods of oscillation from a few minutes to several hours depending on the size of the basin and the nature of the driving force. Seiching action can be initiated by local change in atmospheric pressure (aided by winds), tidal currents, or an earthquake and may continue for days. Loch Tay, a lake in Scotland, is about 14 miles long and 1 mile wide at its broadest part. Unusual seiche action occurred in 1784. William Cruch of Edinburgh

Figure 12

Tsunami of April 1, 1946, inflicting severe damage to coastal buildings in Hawaii. Man in lower left is indicated by arrow. The enormous sea wave was initiated by an earthquake in the Aleutian Trench south of Alaska.
[Courtesy of NOAA/EDS]

described what he saw in a published letter republished by Stewart.

> Upon the 12th of September, 1784, a very extraordinary phenomenon was observed at Loch Tay. The air was perfectly calm, not a breath of wind stirring. About six o'clock in the morning, the water at the east end of the loch ebbed about 300 feet and left the channel dry. It gradually accumulated and rolled on about 300 feet further to the westward, when it met a similar wave rolling in a contrary direction. When the waves met, they rose to a perpendicular height of five or six feet, producing a white foam on the top. The water then took a lateral direction southward, rushing to the shore, and rising upon it four feet beyond the highest watermark. It then returned, and continued to ebb and flow every seven minutes for two hours, the waves gradually diminishing every time they reached the shore, until the whole was quiescent. During the whole of that week, at a later hour in the morning, there was the same appearance, but not with such violence.

The seiche was probably generated by earthquake surface waves.

144 Spaeth, M. G., and Berkman, S. C., 1967, The tsunami of March 28, 1964, as recorded at tide stations: ESSA Technical Report, Coast and Geodetic Survey Technical Bulletin No. 33, 86 pp.

The Alaskan earthquake of March 27, 1964 caused tsunami waves in the Pacific Ocean which reached heights of up to 30 feet (9 meters) at the coast inflicting severe damage in Alaska, Oregon and California. The earthquake surface wave stimulated seiche waves as high as 5 feet (1.5 meters) along the U.S. coast of the Gulf of Mexico. Seiches caused minor damage in rivers, harbors, channels, lakes, and swimming pools in Texas and Louisiana.

45 Yokoyama, I., 1978, The tsunami caused by the prehistoric eruption of Thera, in Thera and the Aegean World I: Second International Scientific Congress, Santorini, Greece, pp. 277-283.

The explosion of the Aegean island volcano Thera (Santorini) in approximately 1500 B.C. generated an enormous tsunami. Sea-borne pumice deposits on the island of Anaphi (25 kilometers east of Thera) and on the coast near Tel Aviv, Israel (1000 kilometers southeast of Thera) indicate tsunami run-up heights of 40 to 50 meters at Anaphi and 7 meters near Tel Aviv. Yokoyama estimates the run-up height at Thera, where the wave was generated, at 63 meters and the height on the north coast of Crete at 11 meters. The Thera tsunami was significantly larger than the Krakatoa (1883) waves.

146 Kumar, N., and Sanders, J. E., 1976, Characteristics of shoreface storm deposits: modern and ancient examples: Journal of Sedimentary Petrology, vol. 46, pp. 145-162.

The authors begin by admitting the not always impartial attitude which has surrounded catastrophic sedimentary processes:

> ...as the geologic philosophy of what might be termed uniformitarianist catastrophism, continues to gain adherents and respectability, the study of storm deposits in the geologic record is losing much of its former 'instant stigma.'...In the past few years the geologic record of ancient nearshore sediments has begun to be interpreted as products of storms...." (page 146)

They then describe modern storm sediments off Long Island, New York. A three-part sequence is recognized: (a) coarse gravel layer at the base--a lag deposit interpreted to represent maximum storm energy and winnowing away of finer material, (b) finely laminated sand in the middle of the deposit, rapidly deposited under conditions of intense bottom shear as the storm waned, (c) coarser, burrow-mottled sand in the upper part--deposited slowly by fair weather after the storm. Comparable storm-deposited sequences are believed to exist in the geologic record and possible examples are documented. The authors conclude:

> If our interpretation of the shoreface sediments described herein is correct, then we predict that the geologic record will be found to contain numerous examples of ancient shoreface storm deposits and doubtless other kinds of storm deposits not related to shorefaces. In fact, we suggest that the geologic record of the nearshore zone will consist largely of storm deposits and that products of the longer-lasting fair-weather conditions will be found to compose only a minor proportion. (page 159)

Catastrophic processes are suggested to be the most important agent of nearshore sedimentation.

147 Perkins, R. D., and Enos, P., 1968, Hurricane Betsy in the Florida Bahama area-geologic effects and comparison with Hurricane Donna: Journal of Geology, vol. 76, pp. 710-717.

The geologic effects of two violent storms are compared and contrasted. The most destructive was Hurricane Donna of 1960 (see Ball et al., 1967, reference 137).

148 Nelson, C. H., 1982, Modern shallow-water graded sand layers from storm surges, Bering Shelf: a mimic of Bouma sequences and turbidite systems: Journal of Sedimentary Petrology, vol. 52, pp. 537-545.

Graded sand layers analogous to the Bouma turbidite sequence have been found in shallow water in the northern Bering Sea, Alaska. The sand layers are believed to have formed during major storm surges when the increased water depth and pore water pressure caused sediment to liquify and flow distances in excess of 100 kilometers over the ocean floor.

149 Lonsdale, P. and Malfait, B., 1974, Abyssal dunes of foraminiferal sand on the Carnegie Ridge: Geological Society of America Bulletin, vol. 85, pp. 1697-1712.

A field of sand dunes is described at a depth of 2.65 kilometers on the floor of the Pacific Ocean west of Ecuador. The dunes averaging 0.6 meter high are believed to form by mysterious, episodic deep-sea currents with velocity in excess of 30 centimeters per second. The currents are not now operating and the dunes are not presently migrating. Other evidence of rapid deep-sea currents are reported. Deep-sea currents attain velocities of 150 centimeters per second in the Norwegian Sea, more than 100 centimters per second out of the Mediterranean Sea, and more than 50 centimeters per second out of the Red Sea. These currents are a potent force in redistributing ocean sediment. Ocean currents of high velocity appear to be needed to form ancient sandstone cross-bedding (see Freeman and Visher, 1975, reference 159).

50 Houtz, R. E., 1962, The 1953 Suva earthquake and tsunami: Seismological Society of America Bulletin, vol. 52, pp. 1-12.

The magnitude 6.75 Suva earthquake in the Fiji Islands of September 1953 caused underwater landslides which initiated a tsunami and produced turbidity currents. The tsunami, which was estimated to have had waves breaking at up to 50 feet height, dislodged coral blocks up to 10 feet in diameter and cast the hulk of a wrecked vessel onto a reef. The turbidity currents are believed to have traveled more than 30 miles along the ocean floor from the undersea landslide area in order to break and displace submarine cables. One cable was displaced 13,000 feet.

51 Miller, D. J., 1960, Giant waves in Lituya Bay Alaska: United States Geological Survey Professional Paper 354-C, pp. 51-83.

Lituya Bay is a T-shaped tidal inlet 7 miles long and up to 2 miles wide with a maximum depth of 720 feet on the northeast shore of the Gulf of Alaska. The head of the bay along Gilbert

Figure 13

Lituya Bay, Alaska, the site of an enormous rock-fall-generated water wave on July 9, 1958. Forest destruction on the shore of the seven-mile-long bay can be clearly seen. In the upper left, the wave leveled trees at an altitude of 1720 feet above sea level. See Reference 151.

[Courtesy of NOAA/EDS]

Water Catastrophes

Figure 14

Wave-eroded slope on the shore of Spirit Lake north of Mount St. Helens, Washington. The May 18, 1980 avalanche from the volcano's north slope created an enormous water wave in Spirit Lake having a run-up height of up to 260 meters (860 feet) on the north shore of the lake. The hill, which is 300 meters (1000 feet) high, is almost entirely denuded of soil and trees which covered its slope before the eruption. Logs removed by the wave now compose the large floating mat on the lake in the foreground.
[Photo by S. A. Austin]

Inlet is a prominent fault where repeated rockslides have generated enormous water waves. At least four giant waves have occurred since 1853.

In 1958 about 40 million cubic yards of rock, loosened either by displacement on the Fairweather fault or by the accompanying shaking, plunged into Gilbert Inlet from a maximum altitude of about 3,000 feet on the steep northeast wall. This rockslide caused water to surge over the opposite wall of the inlet to a maximum altitude of 1,740 feet, and generated a gravity wave that moved out the bay to the mouth at a speed probably between 97 and 130 miles per hour. Two of three fishing boats in the outer part of the bay were sunk, and two persons were killed. The interpretation that water was primarily responsible for destruction of the forest over a total area of 4 square miles, extending to a maximum altitude of 1,720 feet and as much as 3,600 feet in from the high-tide shoreline, is supported by eyewitness accounts of the survivors, by the writer's field investigation, and by R. L. Wiegel's study of a model of Lituya Bay and his calculations from existing theory and data on wave hydraulics.

Perhaps the most destructive rockslide-generated wave in history was at Shimabara Bay, Japan, a bay 60 miles long and 10 miles wide opening to the East China Sea. On May 21, 1792 during a period of intense earthquakes and volcanic activity, some 700 million cubic yards of soil and rock from a maximum elevation of 1,700 feet slid

1.75 miles down a slope of 10 degrees plunging into the sea along 3 miles of the shore. Three large waves followed in rapid succession with the second and largest wave inundating the land to a maximum height of some 33 feet. Trees 9 feet in diameter were snapped, and buildings were destroyed along 50 miles of the shore. More than 15,000 people were killed.

152 Norrman, J. O., 1970, Trends in post-volcanic development of Surtsey Island. Progress report on geomorphical activity in 1968: Reykjavik, Surtsey Research Society, Surtsey Research Progress Report V, pp. 95-112.

Surtsey, a volcanic island 20 miles off the south coast of Iceland, was built from a depth of 130 meters by 1.1 cubic kilometer of volcanic material erupted between November 1963 and July 1967. Norrman's report describes the morphology of the coast in the summer of 1968--one year after the end of volcanic activity--and documents the changes which occurred in the winter of 1967/68. Erosion was most severe on the southern coast where during that winter the basalt cliff retreated up to 140 meters (average 75 meters). Some 200 million cubic meters of basalt were abraded. By the summer of 1968 extensive boulder terrace beaches had formed on the east and west sides of the island. These terraces were up to 100 meters wide consisting of rounded boulders with minor amounts of sand, gravel and cobbles. On the north side of the island, beach processes had built up a cuspate foreland of coarse sand which extended over 300 meters out from the original coast. If a geologist was invited to inspect Surtsey in 1968 without knowledge of its history, he might assign an "age" to it of hundreds of years even though it was less than five years old!

153 Gage, M., 1970, **The tempo of geomorphic change:** Journal of Geology, vol. 78, pp. 619-625.

The cumulative effects of gradual processes and the singular effects of catastrophic processes are contrasted in an attempt to appreciate patterns of landscape change through time. Gage notes that some landforms which appear to require a long time to develop may form rapidly. He describes the Waiho River of New Zealand, which during a single, high-intensity rainstorm in 1965, deposited about 70 feet of sediment on its bed over several miles, then, during succeeding weeks, downcut and eroded its bank to produce a sequence of 10-foot-high terraces. The terraces were quickly colonized by plants and soon acquired a false aspect of antiquity.

154 Kloosterman, J. B., 1976, Overnight valley formation in Sao Nicolau: Catastrophist Geology, vol. 1, no. 2, pp. 44-45

On the night of June 8, 1974, a rain storm in southern Brazil eroded a valley 5 meters deep, 15 meters wide, and 500 meters long in a gently sloping field which before the storm possessed only a small gully. The valley is believed to have formed in less than five minutes due to intense rainfall.

155 Emiliani, C., Gartner, S., Lidz, B., Eldridge, K., Elvey, D. K., Huang, T. C., Stipp, J. J., and Swanson, M. F., 1975, Paleoclimatological analysis of late Quaternary cores from the northeastern Gulf of Mexico: Science, vol. 189, pp. 1083-1088.

A core of deep-sea sediment from the Gulf of Mexico southeast of the Mississippi River delta was analyzed for oxygen isotopes. Foraminiferal shells above a depth of 150 centimeters have significantly less oxygen-18 than the shells in the sediment below a depth of 150 centimeters. The authors believe that the core records an episode of abrupt freshening and warming of the Gulf of Mexico by the addition of Mississippi flood water from rapidly melting glaciers. The

Figure 15

Rapid gulley erosion following the 1980 eruptions of Mount St. Helens. Pyroclastic flow deposits which formed on May 18, 1980, were eroded to a depth of 18 meters (60 feet) within several months following the eruptions. Man in foreground provides scale.
[Photo by S. A. Austin]

time of this flooding is placed at 9,600 B.C. by the authors which coincides with the date of the Flood given by Plato. The flooding and associated sea-level rise are thought by the authors to be a possible explanation of deluge legends from several continents.

Water Catastrophes and the Origin of Sedimentary Rocks

156 Roth, A. A., 1975, Turbidites: Origins, vol. 2, pp. 106, 107.

> Since the advent of the turbidite concept 25 years ago, there has been a significant revolution in the interpretation of a large number of sedimentary deposits. Tens of thousands of graded beds piled upon each other, which were previously interpreted as being slowly deposited in shallow water, are now interpreted as the result of turbidity flows. Even the interturbidite layer, which consists of sediments found between some turbidites, is occasionally interpreted as the result of rapid deposition. This new concept indicates that some events in the past history of the earth may have proceeded much more rapidly than was previously believed. (page 107)

157 Stanley, D. J., 1968, Graded bedding--sole marking--graywacke assemblage and related sedimentary structures in some Carboniferous flood deposits, eastern Massachusetts, in Klein, G., ed., Late Paleozoic and Mesozoic continental sedimentation northeastern North America: Geological Society of America, Special Paper 106, pp. 211-239.

> Floods and turbidity currents have much in common: (1) the flows appear suddenly in a foreign environment; (2) these flows are characterized by high discharge, velocity, and load; (3) the sediment load contains a wide range of size grades; (4) currents are able to erode the bottom and remove the scoured

material; (5) the sediment load is released progressively as the intensity of the flow decreases away from the source. The resulting beds are graded vertically and probably also laterally. Similar depositional processes active in environments as different from each other as continental flood plains and marine basins can produce similar assemblages of sedimentary properties. (page 212)

158 Goldring, R., and Bridges, P., 1973, Sublittoral sheet sandstones: Journal of Sedimentary Petrology, vol. 43, pp. 736-747.

Thin, fine grained sheet sandstones are common in ancient sandy shelf facies. Typically 5-70 cm thick, individual beds exhibit a suite of sedimentary structures: undulose to flat sole, plane or low angle lamination, and wave-rippled tops. The upper part of beds is often bioturbated and/or penecontemporaneously eroded. Two associations are distinguished: a shoreline association and an open shelf association. The first comprises amalgamated beds, the second, amalgamated beds or, more typically, sandstones interbedded with mudstones. Storms, tsunamis, floods, tides, rips and turbidity currents are possible causes. The only recent analogues seem to be the comparatively thin storm generated units described from the North Sea and Texas coast. (page 736)

59 Freeman, W. E., and Visher, G. S., 1975, Stratigraphic analysis of the Navajo Sandstone: Journal of Sedimentary Petrology, vol. 45, pp. 651-668.

The giant cross-bedding in the Navajo Sandstone of Utah is presently a matter of dispute. Some geologists believe it was deposited by wind in a desert environment, while others claim it was deposited by water in a marine environment. Freeman and Visher present textural and structural evidence for water deposition. Brand (1979, reference 160) suggested water origin of cross-bedding in the Coconino Sandstone of the Grand Canyon. To deposit cross-bedding in water requires violent sheet-flow of sediment over giant sand waves. Although such conditions occur in tidal current environments, the thickness and extent of Navajo Sandstone would indicate more extensive strong currents in deeper water. Localized, high velocity deep-sea currents are known (see Lonsdale and Malfait, 1974, reference 149) but not currents on so extensive a scale. Must larger scale currents be postulated for ancient sand deposits?

Figure 16

Giant cross bedding in Navajo Sandstone in Zion National Park, Utah. The origin of the dipping sandstone strata within the horizontal strata has recently been disputed. If deposited by water, the cross bedding would require deep water currents flowing from left to right. Man in lower right corner indicates scale. See Reference 159.
[Photo by S. A. Austin]

160 Brand, L., 1979, Field and laboratory studies on the Coconino Sandstone (Permian) vertebrate footprints and their paleoecological implications: Palaeogeography, Palaeoclimatology, Palaeoecology, vol. 28, pp. 25-38.

Study of fossil footprints in the Coconino Sandstone of the Grand Canyon leads to the conclusion that the footprints were made while the sand was under water. The thick, cross-bedded sandstone may not represent wind-deposited dunes, as most geologists have supposed, but water-deposited sand. If Brand's interpretation is correct, the cross-bedding would require violent sheet flow of water over enormous underwater dunes. Water origin of Navajo Sandstone cross-bedding was suggested by Freeman and Visher (1975, reference 159).

161 Austin, S. A., 1979, Evidence for marine origin of widespread carbonaceous shale partings in the Kentucky No. 12 coal bed (Middle Pennsylvanian) of western Kentucky: Geological Society of America Abstracts with Programs, vol. 11, pp. 381-382.

A coal bed, formed from terrestrial trees, is shown to contain thin marine shale layers called "partings."

The Kentucky No. 12 coal bed of western Kentucky contains a complex sequence of carbonaceous shale partings which is unlike any reported in the literature. In Hopkins, Muhlenberg, and part of Ohio counties, the coal has eight thin carbonaceous shale partings, six of which extended over an area exceeding 1,500 sq. km. As the coal thickens southward from less than 1 cm to over 2 m, the partings split off from the overlying shaly coquina and dark gray mudstone, and pass into the coal bed. At their point of entry into the top of the coal bed, the partings are rich in clay and contain marine fossils. At a distance of over 10 km south of their point of entry, partings usually are deficient in clay, richer in coaly materials, and lack marine fossils. Some partings grade southward into bony coal bands and fusain clast conglomerate. Partings have high illite and quartz, low kaolinite and montmorillonite, normal titanium, and in other ways resemble marine shales above the coal. Partings are rich in inertinite macerals (inertodetrinite, fusinite, and secretion sclerotinite) and have a significant amount of detrital vitrinite. The data is inconsistent with the theory that the partings formed by leaching of thin volcanic ash layers, and at odds with the notion that partings were synthesized in situ by weathering of the peat substrate. River flooding can be discounted as the mechanism for distributing clay because no fluvial channel system was found in association with the coal. Partings in the Kentucky No. 12 coal bed formed by marine flooding of the surface of peat deposition.

162 Stokes, W. L., 1950, Pediment concept applied to Shinarump and similar conglomerates: Geological Society of America Bulletin, vol. 61, pp. 91-98.

The Shinarump conglomerate, a widespread Triassic pebble conglomerate blanket in the southwestern United States, has amazing areal extent. Stokes says:

> The Shinarump conglomerate and recognized equivalents outcrop over wide areas in northern Arizona and southern Utah and less extensively in central, northern and western Utah, western Colorado, northwestern New Mexico, southeastern Nevada, and southwestern Idaho. Its total original areal extent exceeded 125,000 square miles. It is usually less than 50 feet thick but may locally reach 300 feet. (page 91)

Stokes suggests that the conglomerate is the gravel covering of a broad, sloping erosional surface (pediment) flanking a mountain range. A surface so large does not exist today and would require erosion by sheet flooding on a massive scale without channeling and canyon erosion. Might other catastrophic explanations be devised to account for such an unusual feature?

163 Bentor, Y. K., 1980, Phosphorites--the unsolved problems, in Bentor, Y. K., ed., Marine phosphorites--geochemistry, occurrence, genesis: Tulsa, Society of Economic Paleontologists and Mineralogists, Special Publication No. 29, pp. 3-18.

Phosphorites are sedimentary rocks rich in phosphorous. Most geologists regard phosphorite as containing more than about 15 percent P_2O_5 that is of marine origin. Phosphorous is a nutrient in the modern oceans and the element is largely recycled, not deposited. Recent phosphorites form by diagenetic processes after burial, but many ancient examples are of primary origin, formed at the sediment water interface, before burial. In his discussion of the subject "Is the present the key to the past?" Bentor says:

> It seems to follow that, as far as phosphorite formation is concerned, a simplistic actualism does not work. Phosphorite deposition in the past probably took place in a number of ways, only one of which is operative at the present time. The study of alternative mechanisms, in relation to conditions in the ocean, different from the present ones, is an important task for future phosphorite research. (page 13)

64 Carozzi, A. V., and Gerber, M. S., 1978, Late Paleozoic tornadoes and synsedimentary brecciation of chert nodules: Catastrophist Geology, vol. 3, no. 2, pp. 10-18.

A well-exposed section of crinoidal calcarenites of the Lower Burlington near Hannibal, Missouri, USA, displays an unusual occurrence of chert nodules penecontemporaneously brecciated by a short-lived and localized high-energy event interpreted as the touchdown of a tornado system. This example of tempestite demonstrates the syngenetic generation of completely indurated and fragile chert nodules within an unconsolidated carbonate sediment and stresses the fact that only events of catastrophic nature and penecontemporaneous with sedimentation can provide such a demonstration.

Other instances of similar conditions are reviewed and compared with the investigated example. The early origin of cherts is more widespread than generally assumed and a complete reconsideration of the problem of chertification is required. (page 10)

Catastrophic Burial in Fossilization

65 Dunbar, C. O., and Rodgers, J., 1957, Principles of stratigraphy: New York, John Wiley, 356 pp.

> In the coal measures of Nova Scotia, for example, the stumps and trunks of many trees, are preserved standing upright as they grew, clearly having been buried before they had time to fall or rot away. Here sediment certainly accumulated to a depth of many feet within a few years. In other formations where articulated skeletons of large animals are preserved, the sediment must have covered them within a few days at the most. Abundant fossil shells likewise indicate rapid burial, for if shells are long exposed on the sea floor they suffer abrasion or corrosion and are overgrown by sessile organisms or perforated by boring animals. At the rate of deposition postulated by Schuchert, 1,000 years, more or less, would have been required to bury a shell 5 inches in diameter. With very local exceptions fossil shells show no evidence of such long exposure. Evidently then, either our estimates of geologic time are grossly exaggerated, or else most of the elapsed time is not represented in any given section by sedimentary deposits. (page 128, copyright 1957 by John Wiley & Sons, Inc.)

6 Obruchev, V. A., 1977, Fossil cemeteries: Catastrophist Geology, vol. 2, no. 2, pp. 2 and 3.

This reprinted section from Obruchev's book <u>Fundamentals of Geology</u> (1959) describes the bone breccia near Agate, Sioux County, Nebraska.

An even greater cemetery, but of the Tertiary period, was discovered in Carnegie Hill and University Hill in the State of Nebraska. Scores of thousands of skeletons of Rhinoceratidae--Diceras, Moropus and Dinoceras, are buried here in a layer only 15-65 centimeters thick. A slab cut out of this layer and measuring 1.65 x 2 m. contains 22 skulls of Diceras and an enormous mass of its bones in a chaotic mixture. According to available figures 164,000 bones belonging to 820 skeletons of rhinoceroses have already been extracted, most of these bones from skeletons of Diceras. Numerous skeletons of a small antelope-like camel were found in two layers of a neighbouring hill. All the bones are very well preserved and exhibit no marks of teeth of predatory animals or rodents. This shows that the corpses did not stay on the surface very long and were buried very soon. So extensive an accumulation of remains of herbivorous animals of few species in one place can be explained only by a catastrophe which rapidly destroyed whole herds of them. (page 3)

167 Moore, R. C., 1958, Introduction to historical geology: New York, McGraw-Hill, 2nd ed., 656 pp.

The Petrified Forest National Park of Arizona contains thousands of fossil agate and chalcedony logs which have been floated from some unknown location, deposited in layers covering a vast area, and then petrified. Moore says:

> There lie thousands of fossilized logs, many of them broken up into short segments, others complete and unbroken...The average diameter of the logs is 3 to 4 feet, and the length 60 to 80 feet. Some logs 7 feet in greatest diameter and 125 feet long have been observed. None are standing in position of growth but, with branches stripped, lie scattered about as though floated by running water until stranded and subsequently buried in the places where they are now found. The original forests may have been scores of miles distant. The cell structure and fibers have been almost perfectly preserved by molecular replacement of silica.... (pages 401-402)

68 Broadhurst, F. M., 1964, Some aspects of the paleoecology of non-marine faunas and rates of sedimentation in the Lancashire coal measures: American Journal of Science, vol. 262, pp. 858-869.

Not infrequently, large fossils of plants and animals are found to penetrate several strata. Upright fossil trees known as "kettles" or "polystrate trees" may extend through tens of feet of strata, requiring that the sedimentation occurred rapidly before the trees could rot and fall over. Broadhurst describes trees in Lancashire, England:

> In 1959 Broadhurst and Magraw described a fossilized tree, in position of growth, from the Coal Measures at Blackrod near Wigan in

Lancashire. This tree was preserved as a cast, and the evidence available suggested that the cast was at least 38 feet in height. The original tree must have been surrounded and buried by sediment which was compacted before the bulk of the tree decomposed, so that the cavity vacated by the trunk could be occupied by new sediment which formed the cast. This implies a rapid rate of sedimentation around the original tree.... It is clear that trees in position of growth are far from being rare in Lancashire (Teichmuller, 1956, reaches the same conclusion for similar trees in the Rhein-Westfalen Coal Measures), and presumably in all cases there must have been a rapid rate of sedimentation. (pages 865-866)

169 Ballance, P. F., Gregory, M. R., and Gibson, G. W., 1981, Coconuts in Miocene turbidites in New Zealand: possible evidence for tsunami origin of some turbidity currents: Geology, vol. 9, pp. 592-595.

Turbidites are rock strata which are believed to have been deposited from moving, subaqueous, bottom-flowing suspension of sediment and water. A tsunami provides an excellent mechanism for initially creating the suspension and for mixing things from different environments. The authors of the paper found fossil coconuts in deep marine turbidites:

> We think it most probable that the coconuts had lain on the ground long enough for the meat

to rot, when they were caught up in a tsunami and swept offshore in the off-surge, as envisaged by Coleman (1978), along with lots of other plant debris, derived foraminifera and immature, lithic sediment. The latter had not been exposed on a beach for any significant length of time and may have been derived directly from alluvial deposits. It would have been necessary for the off-surge to be channeled directly into a canyon, giving no opportunity for the buoyant coconuts to escape. It was then transformed into a turbidity current. (page 594)

70 Mamay, S. H., and Yochelson, E. L., 1962, Occurrence and significance of marine animal remains in American coal balls: United States Geological Survey Professional Paper 354-I, pp. 193-224.

Coal balls are rounded, mineralized masses of plant fossils found within coal strata. Mamay and Yochelson describe the textural and paleontological characteristics of coal balls from the Midwest. The mixture of terrestrial plants with marine shell material and the concentric structure of the masses requires unusual sedimentary processes:

> The essential feature of the primarily clastic coal balls is the exotic nature of some of their fossil content, which is foreign to the immediate environment of coal deposition. In no sense are the clastic coal balls a normal part of the

sedimentary cyclothem characteristic of Pennsylvanian sedimentation. We believe that their presence within a coal seam along with the more usual, concretionary type of plant-containing coal balls is attributable to unusual, probably catastrophic and transitory means of landward redeposition, such as might be effected by violent storm wave or tidal wave action. (page 195)

171 Jordan, D. S., 1920, A Miocene catastrophe: Natural History, vol. 20, pp. 18-20.

A thin layer of diatomaceous earth near Lompoc, California is estimated to contain one billion fossil herring on a bedding surface covering an area of four square miles. Catastrophic kill is postulated.

172 Morris, S. C., 1979, Burgess Shale, in Fairbridge, R. W., and Jablonski, D., eds., The encyclopedia of paleontology: Stroudsburg, Dowden, Hutchinson & Ross, pp. 153-160.

Soft-bodied fossils from the Middle Cambrian Burgess Shale of British Columbia often preserve muscles, gut, and nerve cords of marine organisms. The exquisite preservation is due to rapid burial. The most famous Burgess fossils are trilobites.

73 Ball, S. M., 1971, The Westphalia Limestone of the northern midcontinent: a possible ancient storm deposit: Journal of Sedimentary Petrology, vol. 41, pp. 217-232.

The Westphalia Limestone, a one-foot-thick layer of limestone occurring in Oklahoma, Kansas and Missouri, contains an estimated 20 trillion small wheat-grain-shaped invertebrate fossils called fusulinids. The amassment of so many fusulinids without much other sediment is considered by Ball to require erosion, current transportation, sorting, and deposition on a vast scale. Ball supposes that the Westphalia Limestone is an ancient storm deposit:

> Petrologic characteristics such as mixed faunas, abrasion and polish of grains, grading in the form of bands of larger invertebrates in the fusulinid sandstone, lime mud rip-clasts as much as 7.5 inches in long dimension, shale pebbles, local current lineation, and the concentration of trillions of fusulinids along with many other invertebrate grain fragments within the fusulinid lime packstone to grainstone are believed to record a current-transported sediment load derived from a shallow seafloor area estimated very conservatively at 1450 square miles. (page 217)

Ancient Geomorphic Features as Evidence of Water Catastrophes

74 Bretz, J H., 1969, The Lake Missoula floods and the Channeled Scabland: Journal of Geology, vol. 77, pp. 505-543.

This paper summarizes Bretz's evidences that an enormous plexus of stream channels occupying 40,000 square kilometers in eastern Washington State formed by catastrophic flooding from the rapid breaching of an ice-dammed lake in Montana. Evidence that the flooding was hundreds of feet deep comes from (1) boulder and gravel bars more than 100 feet high in the middle of channels, (2) subfluvial cataract cliffs, alcoves, and plunge pools hundreds of feet high, (3) silt layers deposited high on slopes by backflooding in valleys tributary to the scablands complex, and (4) the delta at Portland. The first of Bretz's papers on the Channeled Scabland appeared in 1923. His hypothesis was the focus of controversy for 40 years. Geologists reacted by saying: "this heresy must be gently but firmly stamped out." Baker (1978, reference 24) describes the historical development and confirmation of the Spokane flood hypothesis. It is a lesson in why scientists need to let the evidence speak. Bretz received the Penrose Medal, the highest award of the Geological Society of America (see Anonymous, 1980, reference 26).

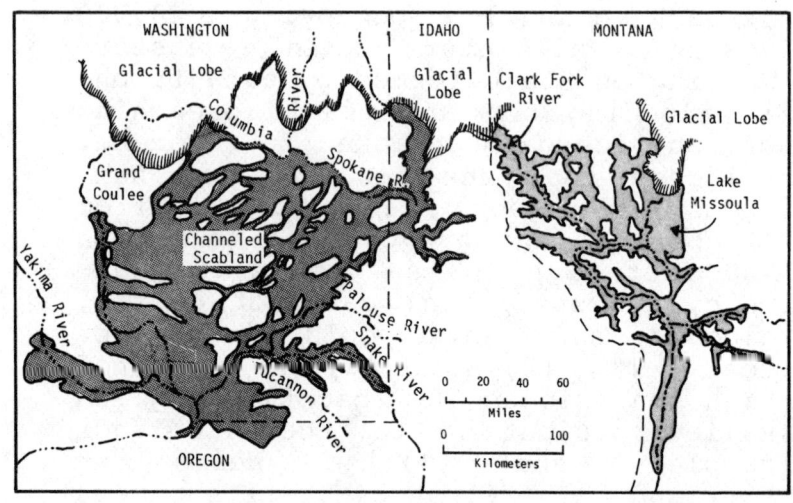

Figure 17

Area devastated by the Spokane Flood. The prehistoric breaching of a glacier ice dam in northern Idaho allowed Lake Missoula in Montana to catastrophically flood Washington State with 500 cubic miles of water (half the present volume of Lake Michigan) producing the Channeled Scablands of eastern Washington. See evidence summarized in Reference 174.

[Map after Baker, Reference 24]

75 Malde, H. E., 1968, The catastrophic Late Pleistocene Bonneville Flood in the Snake River Plain, Idaho: United States Geological Survey Professional Paper 596, 52 p.

Colossal features of erosion and deposition were produced by Pleistocene overflow of Lake Bonneville (the precursor of the Great Salt Lake) onto the Snake River Plain. Evidences of the massive flood are found in abandoned channels, large spillways and cataracts, extensive tracs of scablands, and enormous bars of boulders. Malde estimates the flood volume as 380 cubic miles, the depth of the flood up to 400 feet, the velocity up to 24 feet per second, and the discharge rate of up to 17 million cubic feet per second (0.4 cubic mile per hour).

76 Kehew, A. E., 1982, Catastrophic flood hypothesis for the origin of the Souris spillway, Saskatchewan and North Dakota: Geological Society of America Bulletin, vol. 93, pp. 1051-1058.

Glacial Lake Regina in Saskatchewan drained catastrophically over North Dakota. The Souris spillway is a split-level, water-eroded surface 5 to 10 kilometers wide in Saskatchewan presently occupied by an underfit stream with small drainage basin.

Further evidence of a huge flood is found in North Dakota where an eroded plexus of anastomosing channels, including huge point bars, occur between streamlined erosional remnants. Peak discharge may have been 100,000 cubic meters per second or more. The Souris spillway is not unique:

> The prairie states and provinces of central North America contain many segments of spillway channels similar in size and over-all appearance to the Souris spillway.... If the catastrophic flood hypothesis is supported by the study of other spillways, a major process responsible for drainage diversion and development during continental ice-sheet retreat will have been recognized. (page 1058)

177 Smith, H. T. V., 1965, Anomalous erosional topography in Victoria Land, Antarctica: Science, vol. 148, pp. 941, 942.

An 18-square-kilometer surface in Victoria Land, Antarctica possesses an anastomozing complex of fluvial erosion channels believed to have been carved by rapid melting of the ice cap.

178 Garner, H. F., 1974, The origin of landscapes, a synthesis of geomorphology: New York, Oxford University Press, 734 pp.

Garner is very critical of uniformitarian geology. He says, "the present is like the past as a man is like an amoeba" (page 39). As evidence that modern processes are discordant with ancient processes, Garner describes many relict landforms concluding "the vast majority of the world's landforms are relicts" (page 38). Among the most unusual relict landforms described by Garner are "relict channel labyrinths," enormous dry steam beds which once carried surges of flood waters. These include channels along the Mississippi River in eastern Missouri, along the Ohio River in southern Illinois, in the central Sahara south of Tibisti, in the sculptured terrain of Wright Dry Valley, Antarctica, and in the scabland of eastern Washington State (see pages 440, 441, 512-515).

79 Howard, R. H., 1979, The Mississippian-Pennsylvanian unconformity in the Illinois Basin--old and new thinking, in Palmer, J. E. and Dutcher, R. R., eds., Depositional and structural history of the Pennsylvanian System of the Illinos Basin, part 2: invited papers: Urbana, Illinois State Geological Survey, pp. 34-43.

The conventional idea on the growth of a stream drainage is that the stream lengthens its channel by eroding headward producing a dendritic pattern. The Missippian-Pennsylvanian unconformity in Illinois, Indiana, and Kentucky, a buried surface of erosion, does not possess a <u>dendritic</u> pattern, but is essentially <u>linear</u> with complexly braided and anastomosing channels. Howard compares the erosional pattern on the unconformity to the Channeled Scabland complex of eastern Washington State which was eroded by catastrophic flooding (see Bretz, 1969, reference 174).

180 Olson, W. S., 1966, Origin of the Cambrian-Precambrian unconformity: American Scientist, vol. 54, pp. 458-464.

The Moon is believed to have been captured by the earth just prior to the Cambrian Period. The close approach of the Moon would have produced catastrophic tides, erosion on the world-wide Cambrian-Precambrian unconformity, and deposition of upper Precambrian boulder beds, some of which have been regarded as glacial till by other geologists.

81 Hunt, C. W., 1977, Catastrophic termination of the last Wisconsin ice advance, observations in Alberta and Idaho: Bulletin of Canadian Petroleum Geology, vol. 25, pp. 456-467.

The gravitational encounter between earth and a passing celestial body is believed to have produced giant oceanic tides as much as 5,000 feet (1,700 meters) high causing flooding of the interior of North America. Evidence of extensive oceanic flooding is suggested by distinctive erratic boulders and unusual silt layers in Alberta.

82 Hsü, K. J., 1972, When the Mediterranean dried up: Scientific American, vol. 227, pp. 27-36.

An extensive layer of Miocene anhydrite (calcium sulfate) on the bottom of the Mediterranean Sea is believed to indicate that almost the entire body of water evaporated at an earlier time. The dry basin (2000 miles long, 600 miles wide and 2 miles deep) is believed to have refilled catastrophically with a million cubic miles of water bringing its level even with the Atlantic Ocean. If the depression received 1,000 cubic miles per year (30 million gallons per second) through the Straits of Gibraltar, evaporation loss would be balanced.

Hsü believes the enormous basin filled at ten times this rate. Such a topographic depression would make the Grand Canyon appear like a bath tub, and Niagara Falls like a dripping faucet! An alternate theory might be that a deep-water body was charged with hot, volcanic brines (see Sozansky, 1973, reference 107) so that evaporation is unnecessary.

183 Dury, G. H., 1968, Streams--underfit, in Fairbridge, R. W., ed., The encyclopedia of geomorphology: New York, Reinhold, pp. 1070-1071.

"Underfit streams" are streams which have had a drastic reduction in discharge and which are now too small for the valleys that contain them. They show that there was significantly higher rate of water flow and sediment transport in the past. Dury recognizes the "continent-wide distribution of underfit streams."

184 Dury, G. H., 1965, Theoretical implications of underfit streams: United States Geological Survey Professional Paper 452-C, pp. C1-C43.

The wavelength of a meandering stream varies with the square root of the water discharged through the channel, and shows that many streams are underfit. Dury argues that streams frequently had 20 to 60 times their present discharge in the past.

Physics of Asteroid Impact with Water

85 Gault, D. E. and Wedekind, J. A., 1978, Experimental impact "craters" formed in water: gravity scaling realized: EOS, vol. 59, p. 1121.

High speed projectiles shot into water were photographed by a high speed framing camera. The impact initially forms a cavity in the water, which upon collapse, produces a large central peak having a volume of 60 to 70 percent of the maximum cavity volume. The fall of the water column produces the waves. About 3 to 4 percent of the projectile kinetic energy goes into waves. The experimental data have geologic significance:

> These results also are significant to cratering by large meteoritic masses on 3/4 of Earth's surface. Impact generated tsunamis, whether close to shore or in mid ocean, could devastate (and have in the past) coastal regions of the continental land masses.

86 Gault, D. E., Sonett, C. P., and Wedekind, J. A., 1979, Tsunami generation by pelagic planetoid impact: Abstracts of Tenth Lunar and Planetary Science Conference, pp. 422-424.

This paper is a theoretical study of the tsunami which could be generated by the impact of an asteroid in the deep ocean. Assuming impact of different diameter earth-crossing asteroids (Apollo objects) the wave run-up heights are estimated at great distances from the impact point. For a 1.7-kilometer diameter Apollo asteroid with approach speed of 24.6 kilometers per second (55,000 miles per hour) and silicate density of 3.3 grams per cubic centimeter, the wave run-up height at 1000 kilometers from the impact point is 500 meters. Energy in the waves is about 3 to 4 percent of incident asteroid kinetic energy. The authors say:

> Although such potentially catastrophic inundations of islands and continental margins may be rare occurrences, they represent potentially significant terrestrial events, both biologically and geologically, that have probably occurred in the past and could occur in the future. The search and recognition of such events, which may be stored in the geologic record, offers an interesting and rewarding challenge. (page 423)

187 Strelitz, R., 1979, Meteorite impact in the ocean: Proceedings of the Tenth Lunar and Planetary Science Conference, pp. 2799-2813.

This paper is *not* a rigorous study or complete computation of the wave effects that might be caused by oceanic meteorite impact. The bulk of the paper deals with the propagation of waves and with the complexity in describing nonlinear waves. Strelitz does not estimate wave run-up heights. He says:

> Probability arguments would indicate that about 70 percent of the meteorite influx would land in the ocean, yet there is no obvious and compelling evidence for such events, especially for craters that would, on land, be of the scale of tens of kilometers. The most dramatic effect of such a large impact would perhaps be the generation of a large tidal wave. (page 2799)

88 Kranzer, H. C., and Keller, J. B., 1959, Water waves produced by explosions: Journal of Applied Physics, vol. 30, pp. 398-407.

This classic paper uses the linear theory of surface water waves to describe water waves produced by a point impulse of energy displacing the water surface. The calculations have application to tsunamis such as would be generated by an exploding island or impacting asteroid. The analysis confirms the destructive capability of sea waves generated by large explosive processes.

Chapter 7

ATMOSPHERIC CATASTROPHES

Storms dwarf the works of man. A large thunder cloud may contain half a million tons of condensed water and as much energy as a two-megaton hydrogen bomb. A large hurricane may possess more energy than all the nuclear weapons in the world. The extreme, and often catastrophic, effects of weather concern us because the atmosphere is our home. Yet, severe atmospheric effects are not well understood. This chapter describes air waves generated by catastrophic events (references 189 and 193), documents eruptions and explosions of natural gases (references 194 through 198), explores the climatic effects of catastrophic volcanism (references 199 through 204), investigates glaciation as a possible result of atmospheric catastrophe (references 205 through 209), and relates possible changes in atmospheric composition which could result from an atmospheric catastrophe (references 210 through 211).

Air Waves

89 Hunt, J. N., Palmer, R., and Penney, W., 1960, Atmospheric waves caused by large explosions: Philosophical Transactions of the Royal Society of London, vol. 252, pp. 275-315.

Theory and method are presented for estimating the amplitudes of air waves and energies of events involving large explosions in the atmosphere. The energy of the 1908 Tunguska meteorite is believed to be about 4×10^{24} ergs (10 megatons), close to the energy of the largest nuclear blast. The theory presented can be used to estimate the energies of much larger air waves which would accompany the prehistoric explosions of large asteroids or comets indicated by terrestrial impact craters.

90 Whipple, F. J. W., 1930, The great Siberian meteor and the waves, seismic and aerial, which it produced: Royal Meteorological Society Quarterly Journal, vol. 56, pp. 287-304.

British microbarograph records document air waves from the 1908 Tunguska explosion in Siberia. The waves moved at speeds of up to 323 meters per second, had amplitudes up to 160 microbars, periods of over 3 minutes, and were detected by individual microbarographs for 20 minutes duration.

Whipple estimates the energy of the air waves at 3.2×10^{20} ergs. The air waves were very violent near the source of the explosion where they knocked down 80 million trees over an area of 8,000 square kilometers. A man at a distance of 60 kilometers from the blast was thrown 7 meters through the air (see Krinov, 1966, reference 30).

191 Yokoyama, I., 1981, A geophysical interpretation of the 1883 Krakatau eruption: Journal of Volcanology and Geothermal Research, vol. 9, pp. 359-378.

The physical events of the 1883 eruption of Krakatoa are reviewed with special attention to the blast-generated air waves.

192 Bolt, B. A., 1964, Seismic air waves from the great 1964 Alaskan earthquake: Nature, vol. 202, pp. 1095, 1096.

Exceptional atmospheric waves were detected by microbarographs from the March 28, 1964 Alaskan earthquake. At Berkeley, California, a sound wave coupled to the earthquake surface waves arrived 14 minutes after the earthquake, and a pressure wave generated by earth displacement near the epicenter arrived 2 hours 39 minutes after the earthquake.

93 Wexler, H., and Hass, W. A., 1962, Global atmospheric pressure effects of the October 30, 1961, explosion: Journal of Geophysical Research, vol. 67, pp. 3875-3887.

The test explosion of a gigantic 58 megaton, nuclear device by the Soviet Union on October 30, 1961, caused air waves which circled the earth several times. Its air wave is compared to Krakatoa eruption (1883) and the Siberian meteor (1908).

Eruption and Explosion of Gases

94 Gold, T. and Soter, S., 1980, The deep-earth-gas hypothesis: Scientific American, vol. 242, no. 6, pp. 154-161.

Large earthquakes are known to release gases from within the earth sometimes shooting fire out of the ground, causing bodies of water to bubble fiercely, and producing explosions or hissing sounds.

> The flaming phenomenon indicates that the gas erupting during earthquakes is frequently combustible; most likely it is methane or hydrogen. According to newspaper accounts of the Owens Valley earthquake in California in 1872, "immediately following the great shock, men whose veracity is beyond question ... saw sheets of flame on the rocky sides of the Inyo Mountains but a few miles distant. These flames, observed in several places, waved to and fro, apparently clear of the ground, like vast torches; they continued for only a few minutes." The reality of the flames is verified by the physical evidence that is sometimes left behind. For example, during the Sonora earthquake in Mexico in 1887 a number of people reported flames and later burned branches were found overhanging fissures in the ground. (page 158, copyright 1980 by Scientific American, Inc. All rights reserved.)

95 Gold, T., and Soter, S., 1979, Brontides: natural explosive noises: Science, vol. 204, pp. 371-375.

Sudden eruptions of gas from high-pressure sources within the earth can create natural explosions at the earth's surface. The authors quote W. L. Pierce's eyewitness account of the New Madrid earthquake (December 11, 1812).

> During the first four shocks, tremendous and uninterrupted explosions, resembling a discharge of artillery, were heard from the opposite shore .... Wherever the veins [fissures] of the earthquake ran, there was a volcanic discharge of combustible matter to a great height, an incessant rumbling was heard below, and the bed of the river was excessively agitated, whilst the water assumed a turbid and boiling appearance. Near our boat a spout of confined air, breaking its way through the waters, burst forth, and with a loud report discharged mud, sticks, etc., from the river's bed, at least 30 feet above the surface. (page 375)

196 Gold, T., 1979, Terrestrial sources of carbon and earthquake outgassing: Journal of Petroleum Geology, vol. 1, no. 3, pp. 3-19

Gold presents evidence that combustible gases are frequently released through faults during earthquakes. Although most geologists believe that tsunamis are generated by seafloor displacements along faults, Gold argues that volume change due to eruption of gases on the bottom of the ocean provides a better mechanism.

197 McIver, R. D., 1982, Role of naturally occurring gas hydrates in sediment transport: American Association of Petroleum Geologists Bulletin, vol. 66, pp. 789-792.

Gas hydrates are solid lattices of water molecules which are hydrogen-bonded into hollow spheres or oblate spheroids each of which enclose and contain a gas molecule. Although hydrates contain about six water molecules for each gas molecule, they are not true chemical compounds but rather clathrates or inclusion compounds. Methane is the principal gas contained in hydrate which often saturates enormous masses of ocean sediments imparting rigidity to the sediment because the hydrate acts as a solid. One of the important properties of hydrates is that they are subject to decomposition if pressure decreases or temperature increases. When the hydrate decomposes, the fluidity is restored to the water and gas, which abruptly decreases the strength of the sediment and increases the volume. The overlying sediment may be lifted and/or breached, and the less dense, gas-cut mud may break through. Hydrate decomposition could cause mud diapirs, mud volcanoes, mud slides, turbidite flows, and perhaps some of the "Bermuda Triangle" phenomena. A ship encountering a patch of natural gas would sink and an aircraft could experience engine failure.

Intermittent natural gas blowouts from hydrate-associated gas accumulations, therefore, might explain some of the many mysterious disappearances of ships and planes--particularly in areas where deep-sea sediments contain large amounts of gas in the form of hydrate. This may be the circumstance off the southeast coast of the United States, an area noted for numerous disappearances of ships and aircraft. (page 792)

198 Vogt, P. R., 1972, Evidence for global synchronism in mantle plume convection, and possible significance for geology: Nature, vol. 240, pp. 338-342.

Trace-element pollution by gases expelled from basaltic land volcanoes is believed to have caused the extinction event at the Cretaceous/Tertiary boundary. The extinction event is coincident with the eruption of the Deccan Traps of India and the Brito-Arctic flood basalts.

Climatic Effects
of Catastrophic Volcanism

199 Strommel, H., and Strommel, E., 1979, The year without a summer: Scientific American, vol. 240, no. 6, pp. 176-186.

The summer of 1816 was extraordinarily cold in New England, Canada and western Europe. Snow fell in New England in June and crop-killing frosts continued through August. The "year without a summer" is believed to have been caused by the eruption of Mount Tambora on the island of Sumbawa in Indonesia in 1815. Climatologists rank the Tambora eruption as the greatest producer of atmospheric dust in the last four hundred years. The dust injected into the stratosphere was responsible for global cooling. One wonders what the climatic effects were of much larger prehistoric eruptions (e.g., Toba).

200 Williams, H., and McBirney, A. R., 1979, Volcanology: San Francisco, Freeman Cooper & Co., 397 pp.

The eruption of Laki in Iceland and of Asama in Japan, both in 1783, combined to make that year and the two that followed it three of the coldest on record in the northern hemisphere. They caused "dry fogs" over most of Eurasia and North America. In France, the "fogs" were so dense that the sun could not be seen until it had risen 17 degrees above the horizon. (page 362)

201 Wilson, C. J. N., Ambraseys, N. N., Bradley, J., and Walker, G. P. L., 1980, A new date for the Taupo eruption, New Zealand: Nature, vol. 288, pp. 252, 253.

The Taupo eruption on the North Island of New Zealand produced more than 60 cubic kilometers of volcanic products of which 24 cubic kilometers comprises an unusually violent, pumice-fall deposit (Walker, 1980, reference 66). Twenty-two carbon-14 age determinations estimate the date of the eruption at 131 A.D. (± 17 years), but the authors dispute the accuracy of the date. Instead, the authors believe that the stratospheric dust from the explosion produced the unusual atmospheric phenomenon described in Chinese and Roman history at approximately 186 A.D. Chinese literature describes the Sun "red as blood" at an angle of more than 24 degrees above the horizon.

202 Bray, J. R., 1976, Volcanic triggering of glaciation: Nature, vol. 260, pp. 414, 415.

Closely spaced, massive volcanic ash eruptions are believed to have caused the sudden buildup of permanent snow cover in sub-Artic areas. The snow survived summer melting, increased the albedo of the earth, and, by a series of feedback mechanisms, led to triggering the Pleistocene glaciation.

203 Bray, J. R., 1977, Pleistocene volcanism and glacial initiation: Science, vol. 197, pp. 251-254.

Massive volcanic eruptions are believed to have caused global cooling and initiated Pleistocene glaciation.

204 Axelrod, D. I., 1981, Role of volcanism in climate and evolution: Geological Society of America Special Paper 185, 59 p.

Several major episodes of global explosive volcanism are believed to have sharply lowered temperature and caused extinction of organisms.

Glaciation

205 Hoyle, F., and Wickramasinghe, C., 1978, Comets, ice ages, and ecological catastrophes: Astrophysics and Space Science, vol. 53, pp. 523-526.

High albedo particles with a total mass of 10^{14} grams added to the earth's upper atmosphere would cause an inverse greenhouse effect shielding the surface of earth from sunlight and allowing infrared (heat) radiation from ground level to escape into space. Such particles could be derived from the close passage of a cometary body. This would have severe impact on the climate which would result in marked cooling at ground level. These particles would cause a decrease in the atmosphere's optical depth of visual light which would dramatically affect photosynthetic processes. Food chains, involving larger animals would be disrupted. An ecocatastrophe would take place within one year of the event. Heat transfer from the ocean to the land would take place by evaporation of ocean water which would then be carried and precipitated onto land at an increased rate. Freezing rains would ensue and accumulation of ice on the order of 100 feet per year would be possible in polar regions. Small cause theories for the ice age are disputed. The onset of an ice age is due to an essentially instantaneous very large perturbation of the climate that deposits extensive sheets of ice on land. Fully blown ice ages can vanish with remarkable speed, in approximately one thousand years. Once the ice begins to melt, both local cooling and reflectivity are progressively reduced. The melting occurs exponentially leading to a rapid disappearance of the ice fields.

Butler, E. J., and Hoyle, F., 1979, On the effects of a sudden change in the albedo of the earth: Astrophysics and Space Science, vol. 60, pp. 505-511.

Cometary dust on the order of 10^{14} grams in the upper atmosphere would have severe impact on the earth's climate. This would increase the albedo of the earth causing a severe temperature gradient within air masses between the land and sea. Temperature differences would be between 30 and 50 degrees Celsius. The land masses would cool rapidly within months or weeks. The ocean, because of its tremendous heat storing capacity, would take much longer. The large temperature gradient would set up a thermodynamic engine effect causing heat to be transferred to the land from the ocean with the atmosphere as a working fluid. Gale-force winds of 100 miles per hour would be expected from the sea throughout its thirty year cooling period. Evaporation rates would be greatly increased along with precipitation. The heat reservoir of the ocean is approximately 9×10^{32} ergs, which is sufficient to evaporate 4×10^{22} cubic centimeters of water. Sea level would fall by 100 meters. If distributed uniformly over the land surface, ice thicknesses would reach 300 meters. If distributed from poleward of 50 degrees latitude, an ice thickness of 900 meters would be expected. This would all take place in a few decades after the injection of the dust. Mammoths must have died rather rapidly during the ice age. The belief that mammoths died from natural causes and then were covered by ice cannot explain the extinction of a whole species. These animals were highly adapted to severe climate. The extinction could be explained by a sudden downfall of several feet of freezing rain. Dr. Clare Friend noted that reindeer that fall in crevasses in Greenland were found to be in an unpleasantly putrified condition no matter how cold the surrounding air. Putrification did not take place within many mammoths indicating that they lost their body heat rapidly, much quicker than cold air could do. Freezing rain that opened up the skin could explain rapid freezing. It is also believed that erratic boulders of Norway and England slid great distances on ice sheets driven by high winds. "The earth could emerge from a short-period incidence of a reflective particle blanket locked into a full blown ice-age."

07 Hibben, F. C., 1946, The lost Americans: New York, Crowell, 198 pp.

> Throughout the Alaskan mucks, too, there is evidence of atmospheric disturbances of unparalleled violence. Mammoth and bison alike were torn and twisted as though by a cosmic hand in godly rage. In one place, we can find the foreleg and shoulder of a mammoth with portions of the flesh and the toenails and the hair still clinging to the blackened bones. Close by is the neck and skull of a bison with the vertebrae clinging together with tendons and ligaments and the chitinous covering of the horns intact. There is no mark of a knife or cutting implement. The animals were simply torn apart and scattered over the landscape like things of straw and string, even though some of them weighed several tons. Mixed with the piles of bones are trees, also twisted and torn and piled in tangled groups; and the whole is covered with the fine sifting muck, then frozen solid.
>
> Storms, too, accompany volcanic disturbances of the proportions indicated here. Differences in temperature and the influence of the cubic miles of ash and pumice thrown into the air by eruptions of this sort might well produce winds and blasts of inconceivable violence. If this is the explanation for the end of all this animal life, the Pleistocene period

was terminated by a very exciting time, indeed. (pages 177, 178)

208 Meier, M. F., and Post, A., 1969, What are glacier surges?: Canadian Journal of Earth Sciences, vol. 6, pp. 807-816.

Many glaciers are known to have both quiescent and active phases of movement. The quiescent phase has long duration and slow movement of ice. The active phase has short duration and rapid flow indicating a remarkable degree of decoupling of the glacier from its bed. Surging glaciers can have rates of ice flow of more than 6 kilometers per year for periods exceeding a year. For brief durations ice has been observed to flow as fast as 5 meters per hour.

209 Wilson, A. T., 1969, The climatic effects of large-scale surges of ice sheets: Canadian Journal of Earth Sciences, vol. 6, pp. 911-915.

The rapid surging of the Antarctic Ice Sheet is believed to have increased the albedo of the earth causing the Pleistocene glaciation. The Antarctic Ice Sheet (albedo = 80%) is believed to have displaced and occupied a large area of the southern ocean (albedo = 8%) around the Antarctic continent.

The huge ice advance would transform 10 million square miles of ocean into light reflecting sea ice and decrease the heat input to the earth by about 4%.

Changes in Atmospheric Composition

210 Turco, R. P., Toon, O. B., Park, C., Whitten, R. C., Pollack, J. B., and Noerdlinger, P., 1981, Tunguska meteor fall of 1908: effects on stratospheric ozone: Science, vol. 214, pp. 19-23.

The explosion of a comet in the atmosphere over the Stony Tunguska River, Siberia on June 30, 1908, is believed to have generated as much as 30 million metric tons of nitric oxide (NO) in the stratosphere and mesophere. Calculations indicate that nitric oxide may have consumed up to 45 percent of the ozone in the Northern Hemisphere by early 1909. Measurements of atmospheric transparency for the years 1909 to 1911 indicate steady ozone recovery from unusually low levels in 1909. The Northern Hemisphere appears to have received double the average normal amount of ultraviolet radiation from 1908 through 1911. There is evidence of increase in total arctic sea ice between 1908 and 1911, a decrease in the annual average surface temperature in the Northern Hemisphere for a decade after 1908, and almost 50 percent decrease in the number of tropical cyclones in the Atlantic Ocean and Caribbean Sea from 1910 to 1915. These unusual weather phenomena also coincide with the large volcanic eruptions of Shtyubelya Sopka (1907) and Katmai (1912), both at high latitudes in the Northern Hemisphere.

211 Talman, F. T., 1908, Notes from the Weather Bureau library: Monthly Weather Review, vol. 36, pp. 218-219.

Anomalously bright night sky glow was reported in Europe on June 30, 1908, but its cause was unknown. The recent analysis indicates it was due to solar illumination of dust particles high in the atmosphere after the explosion of the Tunguska comet (see "Tunguska Explosion" in Chapter 3).

> From many parts of middle and northern Europe and the British Isles come reports of a brilliant illumination of the northern sky during the night of June 30-July 1, 1908, and less conspicuous displays of a similar character on other nights preceding and following that date. Nature (London) of July 9 reports that the whole of the northern part of the sky, from the horizon to an altitude of about 45° and extending to the west, was suffused with a reddish hue, the color varying from pink to an Indian red. Several observers state that it was possible to read fairly small print at midnight without any aid from artificial light.
>
> Ciel et Terre (Brussels) reports that in Belgium the illumination, which extended horizontally over an arc of about 90°, did not rise to more than from 5° to 10° above the horizon, though its reflection extended more or less over the whole sky. It was of an intense golden yellow above and a pronounced red below, presenting somewhat the aspect of the eastern sky a few moments before sunrise. The

region of maximum illumination moved slowly toward the east, apparently following the movement of the sun; at midnight it was due north. (page 219)

Chapter 8

RELATED TOPICS

Other subjects are related to the catastrophic processes described in earlier chapters. This chapter documents rapid lithification and fossilization (references 212 through 220), explores the mystery of extinction in the fossil record (references 221 through 225), suggests earth features which have popularly been attributed to slow formation which, instead, may indicate rapid formation (references 226 through 243), and mentions publications which catalog catastrophic processes (references 244 through 249).

Rapid Lithification and Fossilization

Related Topics

212 Sigleo, A. C., 1978, Organic geochemistry of silicified wood, Petrified Forest National Park, Arizona: Geochimica et Cosmochimica Acta, vol. 42, pp. 1397-1405.

"Petrified wood" (wood which has been mineralized by silica) is probably the most well-known fossil to the general public. Sigleo presents evidence that this fossil can form within a few years given reasonable concentrations of dissolved minerals in percolating ground water. Laboratory experiments are reported where the open spaces in wood have been filled with silica within several months. Partial silicification of wood in the laboratory can be accomplished within 24 hours. Wood immersed in alkaline springs in Yellowstone National Park deposits silica at a rate of 0.1 to 4.0 millimeters per year.

213 Oehler, J. H., 1976, Hydrothermal crystallization of silica gel: Geological Society of America Bulletin, vol. 87, pp. 1143-1152.

The chemical sedimentary rock called "chert" has been synthesized rapidly in the laboratory from silica gel under hydrothermal conditions. At a temperature of 300 degrees Celsius and a pressure of 3 kilobars the gel solidifies to microcrystalline quartz within 25 hours. The same degree of crystallization can be achieved at 3 kilobars in 670 hours at 165 degrees Celsius. Crystals form spherulites averaging 20 microns diameter and resemble relict structures that others have incorrectly attributed to biological activity. Blue-green algae were artificially fossilized by the silica gel. Common natural cherts, which have many characteristics of the synthetic chert, must have formed rapidly by crystallization from colloidal precursors at elevated temperature and pressure.

214 Dubrovo, N. A. Giterman, R. Y., Gorlova, R. N., and Rengarten, N. V., 1982, Upper Quaternary deposits and paleogeography of the region inhabited by the young Kirgilyakh mammoth: International Geology Review, vol. 12, pp. 621-634.

In 1977 a bulldozer driver uncovered the frozen corpse of a young mammoth along Kirgilyakh Creek in the Kolyma River basin near Susuman in the Magadan District of eastern Siberia. This was the world's first discovery of the entire body of a young mammoth. The geological, lithological, palynological and paleobotanical features of the deposit containing the mammoth are described by the authors. The mammoth was found lying on its left side in a lens of ice contained in a 6-meter-thick layer of colluvium 150 meters from the creek channel, buried at a depth of 2 to 2.5 meters below the surface, at a height of 8 to 9 meters above the creek level. The colluvium (containing the pure, colorless ice lens with the mammoth) consists of unsorted, unbedded cobbles, pebbles, sand, silt and clay which forms a widespread layer in the creek valley. The colluvium is overlain by a thin layer of soil and is underlain by unconsolidated alluvial gravel which forms distinct buried terraces. Subaerial origin is postulated for the colluvium because it contains friable rotten rock fragments cemented by ice which could not have been transported any distance by flowing water. A buried soil on a covered terrace immediately below the colluvium contains upright stumps of larch and birch trees in growth position penetrating up into the colluvium requiring rapid accumulation of the colluvium. Analysis of plant fragments, spores and pollen at the level of the mammoth indicates that the colluvium formed, and the mammoth lived, in a tundra-steppe and larch open-forest environment similar to the modern Kolyma River basin. The mechanics of colluvium deposition and the circumstances causing the mammoth to be frozen in the ice lens remain a mystery.

215 Alexandersson, T., 1970, The sedimentary xenoliths from Surtsey: marine sediments lithified on the sea-floor, a preliminary report: Reykjavik, Surtsey Research Society, Surtsey Research Progress Report V, pp. 83-89.

During the volcanic eruption of Surtsey (1963-1967) fossiliferous sedimentary rock fragments (xenoliths) were incorporated in the rising magma and are now found in the tephra on the island. Fossil shells, virtually identical to shells of modern organism living in the sea around Iceland today, argue that the sedimentary rocks are of recent origin. Furthermore, the active volcanic and tectonic setting of Surtsey should prohibit "old" sedimentary rocks from existing on the ocean floor. Because the rock was lithified before the Surtsey eruption (November 14, 1963) and because it was derived from the seafloor where Surtsey stands, the xenolithic material provides evidence of rapid submarine fossilization and lithification while the sediment was in a poorly compacted state at or near the sediment/water interface. This is contrary to the views of many geologists who regard lithification and fossilization as a result of a slow process of deep burial and compaction followed by uplift out of the sea and finally cementation. Alexandersson believes that the sedimentary

Related Topics

xenoliths of Surtsey were lithified recently below the ocean in connection with submarine volcanism.

16 Shinn, E. A., 1969, Submarine lithification of Holocene carbonate sediments in the Persian Gulf: Sedimentology, vol. 12, pp. 109-144.

Large areas of sea floor in the Persian Gulf are composed of carbonate sediment with occasional fragments of pottery and other artifacts cemented by the minerals aragonite and calcite. These recently lithified sediments show that long periods of time and deep burial are not required to form sedimentary rock.

17 Davis, A., and Spackman, W., 1964, The role of the cellulosic and lignitic components of wood in artificial coalification: Fuel, vol. 43, pp. 215-224.

Wood from the bald cypress (Taxodium distichum) was subjected to high temperatures and pressures producing coal vitrinoid substances in less than 10 days. Heat is the principal cause of artificial coalification.

218 Gentry, R. V., Christie, W. H., Smith, D. H., Emergy, J. F., Reynolds, S. A., Walker, R., Cristy, S. S., Gentry, P. A., 1976, Radiohalos in coalified wood: new evidence relating to the time of uranium introduction and coalification: Science, vol. 194, pp. 315-318.

Radiohalos are spherical zones of discoloration around microscopic radioactive mineral grains. The size and structure of radiohalos are determined by the type of radioactive element or elements at the center. Halos of the short half-life emitter ^{210}Po (half-life = 138 days) have been found in coalified wood requiring that the process of coalification must occur rapidly (within a few years) while polonium was present. Halos of the long half life emitter ^{238}U (half-life = 4.5 billion years) from coalified wood do not have significant amounts of daughter element ^{206}Pb indicating that only a few thousand years have elapsed between the time of coalification and the present. These data call into question the slow process of coalification and millions-of-years ages assumed for coalified wood.

219 Gentry, R. V., 1974, Radiohalos in radiochronological and cosmological perspective: Science, vol. 184, pp. 62-66.

The minerals biotite, fluorite and cordierite have been observed to contain polonium radiohalos. Because the half lives of polonium isotopes are measured in periods of a fraction of a year, a slow igneous or hydrothermal process could not have formed these minerals. The radiohalos argue for rapid crystallization.

220 Giannasi, D. E., and Niklas, K. J., 1977, Flavonoid and other chemical constituents of fossil Miocene <u>Celtis</u> and <u>Ulmus</u> (Succor Creek Flora): Science, vol. 197, pp. 765-767.

Green-colored fossil leaves buried in tuff contain flavonoids and chlorophyll derivatives that require rapid burial, low temperature, and neutral pH in order to prevent thermolytic decomposition.

Extinction and the Fossil Record

221 Schindewolf, O. H., 1977, Neocatastrophism?: Catastrophist Geology, vol. 2, no. 2, pp. 9-21.

This article is an English translation of the address to the General Assembly of the German Geological Society in 1961.

> The acceptance of faunal discontinuities in the history of the Earth has lately been, somewhat disparagingly, described as neocatastrophism and represented as a regression to the long-discredited ideas of Cuvier and his time. A reexamination of the issues has yielded the result that at the turning points of the great geological eras, and to a lesser extent at the boundaries of formations, there have occurred fundamental changes in the composition of the animal world, produced by the massive and more or less simultaneous extinction of numerous stocks and the appearance of new ones. This universal phenomenon must be regarded as a reality, and its causal interpretation requires postulating factors that are universally active. An admissible interpretation may be seen in the effects of influxes of high-energy cosmic radiation. (page 9)

222 Ager, D. V., 1976, The nature of the fossil record: Proceedings of the Geological Association, vol. 87, pp. 131-159.

> The fossil record, like the stratigraphical record, is thought to be episodic with long periods of quiescence separated by short

periods of explosive evolution, expropriation and extinctions. Examples are discussed mainly from the bivalves, the cephalopods, the brachiopods and the vertebrates. Terrestrial and extraterrestrial causes are discussed and the balance of the evidence is thought to put the blame firmly on marine transgressions and regressions, controlled by plate movements and mantle plume activity. Conservation, whether it be of species, environments, nations or languages, is doomed, but the author remains cheerful. (page 131)

223 Russell, D. A., 1979, The enigma of the extinction of the dinosaurs: Annual Reviews of Earth and Planetary Science, vol. 7, pp. 163-82.

Russell considers a variety of causes for the extinction of the dinosaurs: trophic effects, marine regressions, temperature changes, volcanism, collisions of comets and large meteorites, increase in ultraviolet or ionizing radiation, supernovae effects, and periodic galactic effects. Catastrophic causes are freely discussed.

224 McLaren, D. J., 1970, Presidential address: time, life, and boundaries: Journal of Paleontology, vol. 44, pp. 801-815.

A well-known paleontologist proposes that Devonian marine organisms became extinct due to a catastrophic flood caused by meteorite impact.

> Presumably on impact with the ocean surface or at a certain depth below the surface, the missile will explode with an enormouse release of energy. In the copious literature on meteorites, impact craters and astroblemes, I have been unable to find a calculation of the energy effect in terms of tidal waves from such an explosion. Dietz (1961) suggests that a giant meteorite falling in the middle of the Atlantic Ocean today would generate a wave twenty thousand feet high. This will do. The effect of such an explosion would certainly spread to all shelf and epicontinental areas connected with the open ocean. The turbulence of the tidal wave and accompanying wind, followed by the gigantic runoff from the land would induce a turbid environment far longer than could be survived by bottom dwelling filter-feeders, and their larvae. The hypothesis of meteoric impact in the ocean explains equally the non-extinction of many other forms of both marine and terrestrial life. (page 812)

225 McLaren, D. J., 1983, Bolides and biostratigraphy: Geological Society of America Bulletin, vol. 94, pp. 313-324.

In his presidential address to the Geological Society of America, McLaren gives reasons why he believes that mass extinctions in the paleontological record may be explained in terms of global catastrophes, the most probable 'cause being the impact of large cosmic bodies.

Features Attributed to Slow Formation Which May Indicate Rapid Formation

226 Lambert, A., and Hsü, K. J., 1979, Non-annual cycles of varve-like sedimentation in Walensee, Switzerland: Sedimentology, vol. 26, pp. 453-461.

Thin, rhythmic silt and clay layers found in lakes are frequently called "varves," with each layer being considered to represent annual repetitions of a slow sedimentary process. Lambert and Hsü present evidence from a Swiss lake that these varve-like layers form rapidly by catastrophic, turbid water underflows. At one location five "varves" formed during a single year.

227 Plummer, P. S., and Gostin, V. A., 1981, Shrinkage cracks: desiccation or synaeresis?: Journal of Sedimentary Petrology, vol. 51, pp. 1147-1156.

Thick sedimentary rock sequences containing shrinkage cracks are frequently claimed to have required repeated wetting and drying of the sediment surface, thereby requiring a lot of time in what would normally appear to be a rapidly deposited sedimentary sequence. Plummer and Gostin urge caution when interpreting "mudcracks" in rocks as evidence of drying.

> Shrinkage cracks can form not only at the sediment-air interface by dessication processes

but also at the sediment-water interface or substratally by synaeresis processes." (page 1147)

28 Fritz, W. J., 1980, Reinterpretation of the depositional environment of the Yellowstone "fossil forests": Geology, vol. 8, pp. 309-313.

Fritz disputes the idea that the petrified vertical tree stumps of the Eocene Lamar River formation of Yellowstone National Park represent in situ burial of multiple forests. Instead, he presents evidence that stumps have been moved in mass, deposited in right-side-up position, and buried by alluvial sediments. These rapid processes are believed to have occurred on the flanks of volcanoes producing a complex series of sedimentary rocks where no "forest" or depositional layer exists that can be traced for any distance, and helps explain mixing of plants from different ecological zones.

29 Rupke, N. A., 1969, Sedimentary evidence for the allochthonous origin of Stigmaria, Carboniferous, Nova Scotia: Geological Society of America Bulletin, vol. 80, pp. 2109-2114.

Stigmaria, the rootlike organ of lycopod trees, has often been cited by geologists as unambiguous proof of growth-in-place of trees within sedimentary strata sequences. Rupke challenges, the *in situ* interpretation of *Stigmaria* and offers four evidences for transportation prior to burial: (1) preferred orientation of the long axes, (2) fragmentation, not attachment to stumps, (3) filling with sediment unlike the enveloping rock, and (4) rapid accumulation of sedimentary beds containing *Stigmaria*.

230 Schultz, L. G., 1958, Petrology of underclays: Geological Society of American Bulletin, vol. 69, pp. 363-402.

Underclays below coal beds are frequently regarded as residual soils slowly formed beneath swamps in which coal-forming flora grew. Schultz disputes the residual soil theory presenting evidence for transportation and flocculation of clay, not weathering *in situ*.

231 Simoneit, B. R. T., and Lonsdale, P. F., 1982, Hydrothermal petroleum in mineralized mounds at the seabed of Guaymas Basin: Nature, vol. 295, pp. 198-202.

Deep ocean springs in the Gulf of California contain patches of "young" petroleum, evidence that oil can form rapidly in natural environments. The springs contain circulating hot water which operates much like a petroleum refinery breaking organic chemicals and accelerating the oil formation process, producing in thousands of years what might appear to require millions of years. Gasoline-range hydrocarbons and asphaltic material sampled at the seabed formed recently from "immature" biological detritus, not, for example, from older "mature" petroleum.

232 Fisher, L. W., 1934, Growth of stalactites: American Mineralogist, vol. 19, pp. 429-431.

Fisher summarizes some of the literature on rapid growth of stalactites and stalagmites. Growth rates of stalactites average about 1.25 centimeters (0.5 inch) yearly with some observed to grow over 7.6 centimeters (3 inches) yearly. Stalagmites observed by Fisher grew 0.6 centimeter (0.25 inch) in height and 0.9 centimeter (0.36 inch) in diameter at the base each year.

233 Blatt, H., Middleton, G., and Murray, R., 1972, Origin of sedimentary rocks: Englewood Cliffs, Prentice-Hall, 634 pp.

The word "reef" has been abused by geologists. Frequently, "reef" simply refers to any limestone body with upward convexity, whether or not it contains a growth framework of sediment-binding marine fossils. These limestone deposits are often considered to require very long periods of time to slowly accumulate, but the growth framework, the proof of slow deposition, is usually lacking. The authors say:

> Carbonate mound-like features or bioherms are well-known in the ancient record. Many of them contain abundant organic remains. Closer inspection of many of these ancient carbonate 'reefs' reveals that they are composed largely of carbonate mud with the larger skeletal particles "floating" within the mud matrix. Conclusive evidence for a rigid organic framework does not exist in most of the ancient carbonate mounds. In this sense they are remarkably different from modern coral-algal reefs. (pages 410, 412)

234 Heckel, P. H., 1972, Possible inorganic origin for stromatactis in calcilutite mounds in the Tully Limestone, Devonian of New York: Journal of Sedimentary Petrology, vol. 42, pp. 7-18.

The flat-bottomed, carbonate, spar-filled structure called "stromatactis" occurring in limestones has been interpreted as the rigid skeleton of a frame-building organism that grew slowly in a reef. Heckel notes the absence of evidence of rigid skeletons and argues that the structure formed rapidly by sediment collapse and subsequent inorganic precipitation in water-filled cavities within fine-grained, limy sediment.

35 Jehl, J. R., 1983, Tufa formation at Mono Lake, California: California Geology, vol. 36, p. 3.

Tufa is the calcium carbonate rock precipitated from mineral-rich solution on the bottom of lakes. Mono Lake in Mono County, California, has large towers of tufa deposited where calcium-bearing springs enter the bed of the lake. A 17-year-old steel drum deposited by the U. S. Navy on the lake bed was observed to have a thickness of 43.2 centimeters of tufa deposited over it indicating an average rate of tufa deposition of 2.54 centimeters (one inch) per year. A 10-meter-high tufa tower around a spring may have formed in as few as 400 years.

36 Kranz, P. N., 1974, Computer simulation of fossil assemblage formation under conditions of anastrophic burial: Journal of Paleontology, vol. 48, pp. 800-808.

Clams have the ability to burrow out of sediment rapidly deposited on top of them. A computer model is used to simulate the assemblage of clams that would die in various "anastrophic" (localized catastrophic) situations. This emphasizes that a sequence of burrowed sedimentary strata appearing to contain the normal occupation burrows inhabited during long periods of time may, instead, represent escape burrows in rapidly deposited sediment.

237 **Twidale, C. R., 1976, On the survival of paleoforms: American Journal of Science, vol. 276, pp. 77-95.**

The fact of modern landscapes containing very old paleoforms (especially paleosurfaces of low relief) is held by Twidale to be inconsistent with some of the commonly espoused models of landscape evolution. Extensive periods of time coupled with relentless slow erosion processes are not regarded as effectively leveling mountainous topography to form plains. Slow processes over extended time may, according to Twidale, increase relief or leave relatively unaffected older landforms. One possible consequence of Twidale's ideas is that the ancient paleoforms preserved for our observation may have been created by ancient processes which operated at a rate, scale or intensity exceeding modern processes.

238 Cotton, C. A., 1968, Relict landforms, <u>in</u> Fairbridge, R. W., ed, The encyclopedia of geomorphology: New York, Reinhold, pp. 936-940.

Relict landforms are surface features created by erosive processes which are no longer acting. Because they are no longer being made, relict landforms present problems for evolutionary and uniformitarian theories of landscape origin which suppose that present landforms are the products of long continued action of processes operating at a rate and scale like today. Instead, relicts indicate discontinuity between modern and ancient processes and climates, and lend themselves most naturally to catastrophic interpretations.

Cotton describes several types of relict landscapes. The Sahara of Africa possesses many indicators of wet climate. Pediments in Spain are not forming now but required excessive rainfall. Streams in Europe are misfit to the valleys that contain them. Sand dunes in western Nebraska cover an area of about 22,000 square miles, but are not moving being vegetated today in the more humid climate. Sea cliffs on Campbell Island and the Auckland Islands south of New Zealand have been partly submerged by sea level rise and are no longer undergoing marine erosion. Cotton cites the work of Budel that practically the whole of Germany and the rest of northwest Europe is a relict landscape remaining from Pleistocene glaciation. Indeed, Budel says that almost everywhere relict landforms occupy a far greater area than the landforms developing by currently acting processes.

239 Schumm, S. A., 1975, Episodic erosion: a modification of the geomorphic cycle, in Melhorn, W. N., and Flemal, R. C., eds., Theory of landform development: London, George Allen & Unwin Ltd., pp. 69-85.

Schumm disputes the notion of William Morris Davis that landscapes slowly evolve as stream gradients and valley floor altitudes change almost imperceptibly through time. Instead, Schumm advocates brief periods of instability and incision of river beds and valley floors separated by long periods of relative stability. Douglas Creek, occupying a 400-square-mile drainage basin in western Colorado, has four unpaired, discontinuous terraces, up to 6.5 meters above the present creek, which have formed since 1882. In the evolutionary view of Davis, the surfaces would be explained as much older formed by shifting of the channel laterally across the valley floor during continuous downcutting. Schumm argues that Douglas Creek shows brief intervals of valley deposition and incision followed by long periods of stability.

240 Born, S. M., and Ritter, D. F., 1970, Modern terrace development near Pyramid Lake, Nevada, and its geologic implications: Geological Society of America Bulletin, vol. 81, pp. 1233-1242.

Six levels of stream terraces along the Truckee River in Nevada formed during a period of only 44 years. Such terraces have usually been assumed to represent long periods of geomorphic equilibrium instead of short periods of rapid change.

241 Smith, P. J., 1977, Ferromanganese deposits: fast, fast, slow: Nature, vol. 265, pp. 582, 583.

Ferromanganese deposits, popularly known as "manganese nodules," occur as nodules, slabs, crusts or grains on the ocean floor. They contain oxide minerals with as much as 55 percent manganese, 35 percent iron, and 2 percent nickel, cobalt and copper. Although the potential economic importance of ferromanganese deposits has encouraged scientific research, some major problems concerning the geologic and chemical processes involved in growth remain to be solved. Smith reviews the literature on the rates of growth of ferromanganese deposits. Some scientists believe that ferromanganese is a rapid precipitate from submarine volcanic exhalations, whereas others suppose that the material grows very slowly from normal seawater with iron and manganese being derived from the continents. Evidence of rapid precipitation comes from a 50-year old naval shell possessing almost 30 millimeters of ferromaganese coating and a World War II shell with a 15 millimeter coating, both discovered on the seafloor. This provides evidence of rapid rates of deposition exceeding one half millimeter per year. Ferromanganese not coating historical objects is more difficult to date. Radioactive dating methods using thorium and uranium do not agree with the rates from historically dated objects, and the radioactive methods often differ with each other, but usually yield very slow deposition rates as low as a few millimeters per million years. This casts doubt on radioactive dating methods used to estimate the age of ferromanganese and causes one to look critically at uniformitarian theories for the origin of these deposits.

242 Cowan, G. A., 1976, A natural fission reactor: Scientific American, vol. 235, no. 1, pp. 36-47.

Six natural, nuclear-fission reactors are believed to have existed in a rich deposit of uranium at Oklo in the Gabon Republic in west Africa.

43 Hall, J. M., and Robinson, P. T., 1979, Deep crustal drilling in the North Atlantic Ocean: Science, vol. 204, pp. 573-586.

The Vine-Matthews hypothesis of sea-floor spreading which was proposed in the early 1960's, suggests that linear magnetic anomalies detected by magnetometers above the mid-oceanic ridges are caused by remnant magnetism resident in oceanic basalt formed by magnetic field reversals as new ocean floor formed continuously over periods of millions of years. The slow process of divergence of crustal plates is now reported as fact by many geology textbooks, and linear magnetic anomalies are featured as prima facie evidence. Hall and Robinson report the findings of the first direct test of the Vine-Matthews hypothesis--deep crustal drilling of oceanic basalts. The anticipated magnetic orientation was not found.

> Where and what is the source of the linear magnetic anomalies on the seafloor? Drilling has shown convincingly that the source of the anomalies does not lie in the upper 600 m of oceanic crust as previously thought. Are the anomalies generated by considerable thick-

nesses of low-intensity material or is there some highly magnetic source layer at greater depth? If so, what constitutes the layer and why is it not tectonically (and thus magnetically) disrupted in the same way as upper layer 2? (page 585)

Catalogs of Catastrophic Processes

244 McWhirter. N., ed., 1982, Guinness book of world records: New York, Bantam Books, 20th ed., 704 pp.

The most energetic earthquake on seismograph records is the Lebu, Chile earthquake of May 22, 1960, measuring 9.5 on the new Kanamori scale (but only fourth place at magnitude 8.3 on the Richter scale) releasing an energy of approximately 1×10^{26} ergs. It is possible that the Chile earthquake was exceeded by the Lisbon, Portugal earthquake of November 1, 1755. The most destructive earthquake in terms of human life occurred in Shensi Province, China on January 23, 1556 when an estimated 830,000 people were killed. Much larger earthquakes must have accompanied the formation of large asteroid impact craters (see Clube and Napier, 1982, reference 134).

The most violent volcanic eruption in history is the Taupo eruption in New Zealand in approximately 186 A.D. (see reference 201). Much larger volcanic eruptions can be documentd from the geologic record (see Heiken, 1979, reference 78).

The highest documented seismic sea wave (tsunami) in the open ocean appeared on April 24, 1971 off Ishigaki Island, Ryukyu Chain having an estimated height of 278 feet. The enormous wave tossed an 850-ton block of coral a distance of 1.3 miles. Much larger sea waves are a very probable result of asteroid impact (see Gault et al., 1979, reference 186) and their sedimentary products appear to await discovery in the geologic record.

45 Frazier, K., 1979, The violent face of nature, severe phenomena and natural disasters: New York, William Morrow & Co., 386 pp.

This book, written in popular style, portrays the most devastating forces of nature which have been experienced by man: thunderstorms, tornadoes, lightning, hail, floods, hurricanes, blizzards, volcanoes, and earthquakes.

> At any given moment, 1,800 thunderstorms are in progress over the earth's surface. Lightning is striking the earth 100 times each second. If the season is late summer, one or more of the some 50 hurricances or typhoons that swirl into existence each year is likely to be moving toward a populated coastline. If the time is late afternoon, the odds are good that a tornado is raking across the American heartland; 600 to 1,000 times a year they do so, and in the prime months they can strike with a frequency of four or more a day. Somewhere at any given moment people's homes or crops are under flood waters. (page 13)

246 Scheidegger, A. E., 1975, Physical aspects of natural catastrophes: New York, Elsevier, 289 pp.

The physical processes that are involved in natural catastrophes are analyzed quantitatively from the standpoint of pure science and engineering. The book includes details concerning earthquakes, volcanic eruptions, accidents on slopes (rockfall, landslide, etc.), snow and ice catastrophes (avalanche, surging of glacier, etc.), water catastrophes (flash flood, tsunami, etc.), and air catastrophes (severe weather).

247 Brongersma-Sanders, M., 1957, Mass mortality in the sea: Geological Society of America Memoir 67, vol. 1, pp. 941-1010.

A large number of historical mass mortality events involving marine organisms are classified and described. The major causes of catastrophic kill of marine organisms are volcanism, earthquake, change in salinity, change in temperature, noxious waterblooms, and lack of oxygen. An example of mass mortality in the sea occurred in March and April 1882 between Cape May and Nantucket on the east coast of the United States where enormous numbers of dead fish appeared on the ocean surface over an area 170 by 25 miles. The cause of this fish kill appears to have been invasion of colder bottom water into the warmer water of the Gulf Stream.

248 Corliss, W. R., 1980, Unknown earth: a handbook of geological enigmas: Glen Arm, Sourcebook Project, 833 pp.

This handbook cites the literature describing some of the most interesting and controversial geologic phenomena. Included are readings on modern and ancient catastrophic processes.

249 Corliss, W. R., 1977, Handbook of unusual natural phenomena: Glen Arm, Sourcebook Project, 542 pp.

This handbook contains an incomparable assemblage of descriptions of rare and fascinating natural phenomena, some of which are beyond explanation by current science. Modern catastrophic geologic processes are described in chapters on "Mysterious Natural Sounds," "Strange Phenomena of Earthquakes," and "Phenomena of the Hydrosphere."

AUTHOR INDEX

Author	Reference No.
Ager, D.V.	3, 222
Albritton, C.C., Jr.	17
Alexandersson, T	215
Alt, D.D.	79
Alvarez, L.W.	44
Alvarez, W.	44
Ambraseys, N.N.	201
Anderson, O.L.	95
Anonymous	26
Asara, F.	44
Austin, S.A.	7, 28, 161
Axelrod, D.I.	204
Bailey, E.B.	114
Bailey, E.H.	104
Bailey, R.A.	77
Baker, V.R.	24
Ball, M.M.	137
Ball, S.M.	173
Ballance, P.F.	169
Barberi, F.	84
Bates, R.L.	6
Bemmelen, R.W. Van	75
Ben-Menahem, A.	29
Bentor, Y.K.	163
Berkman, S.C.	144
Bischoff, J.L.	105
Blatt, H.	233
Boekschoten, G.J.	65
Bogoslovsky, V.	55
Bolt, B.A.	192
Bombolakis, E.G.	120
Bond, A.	64
Born, S.M.	240
Bradley, J.	201

Author	Reference No.
Brand, L.	160
Bray, J.R.	202, 203
Brazo, M.W.	28
Brenner, R.L.	23
Bretz, J H.	174
Bridges, P.	158
Broadhead, G.C.	127
Broadhurst, F.M.	168
Brongersma-Sanders, M.	247
Bronshten, I.A.	31
Browning, J.M.	110
Bullard, F.M.	67
Burns, V.M.	105
Butler, E.J.	206
Calame, O.	52
Carey, S.	63
Carozzi, A.V.	164
Chadwick, A.V.	112
Chapman, C.R.	34
Chinnery, M.A.	131
Choubey, V.D.	88
Christiansen, R.L.	80
Christie, W.H.	218
Chuckuru-Ike, M.	42
Clark, D.H.	53
Clube, S.V.M.	35, 48, 49, 134
Coleman, P.J.	142
Conaghan, P.J.	115
Conybeare, W.D.	11
Coombs, H.A.	90, 91
Corliss, W.R.	248, 249
Cory, H.T.	140
Cotton, C.A.	238
Cowan, G.A.	242
Cristy, S.S.	218
Crutzen, P.J.	54

Author Index

Author	Reference No.
Curray, J.R.	125
Dachille, F.	50
Dalrymple, G.B.	77
Damberger, H.H.	102
Davies, D.K.	23
Davis, A.	217
Davis, D.R.	34
Davis, E.E.	93
Dawson, J.B.	69
Decker, B.	56, 57
Decker, R.	56, 57
Dence, M.R.	36
Dietz, R.S.	38
Doehring, D.O.	139
Donn, W.L.	73
Dubrovo, N.A.	214
Dunbar, C.O.	165
Dury, G.H.	5, 183, 184
Edgecombe, D.R.	115
Eldridge, K.	155
Elvey, D.K.	155
Emergy, J.F.	218
Emiliani, C.	155
Emmel, F.J.	125
Enos, P.	147
Fairbridge, R.W.	74
Federman, A.	63
Fisher, L.W.	232
Fisher, R.V.	72, 86
Foster, H.L.	129
Frazier, K.	245
Freeman, W.E.	159
Friedman, G.M.	21
Fritz, W.J.	228
Gage, M.	153
Ganapathy, R.	47

Author	Reference No.
Garner, H.F.	178
Gartner, S.	155
Gault, D.E.	133, 174, 186
Gentry, P.A.	218
Gentry, R.V.	218, 219
Gerber, M.S.	164
Giannasi, D.E.	220
Gibson, G.W.	169
Giterman, R.Y.	214
Gold, T.	194, 195, 196
Goldring, R.	158
Goles, G.	82
Goodman, N.	18
Gorlova, R.N.	214
Gostin, V.A.	227
Gould, S.J.	16
Gregory, M.R.	169
Grieve, R.A.F.	36, 37
Grinnell, G.	15
Gussow, W.C.	100
Hall, J.M.	243
Hass, W.A.	193
Hayes, M.O.	138
Heckel, P.H.	234
Heiken, G.	78
Helin, E.F.	33
Helz, R.T.	89
Heylmun, E.B.	2
Hibben, F.C.	207
Hooykaas, R.	12
Houtz, R.E.	150
Howard, R.H.	179
Hoyle, F.	205, 206
Hsü, K.J.	27, 51, 109, 182, 226
Huang, T.C.	63, 155
Hulme, G.	71

Author Index

Author	Reference No.
Hunt, C.W.	181
Hunt, J.N.	189
Hutchinson, J.N.	111
Hyndman, D.W.	79
Imboden, O.	62
Innocenti, F.	84
Irvine, T.N.	99
Irwin, W.P.	104
Izett, G.A.	83
Jackson, J.A.	6
Jehl, J.R.	235
Johns, D.R.	113
Jones, D.L.	104
Jordan, D.S.	171
Kanamori, H.	130
Karlstrom, T.N.V.	129
Kastner, M.	105
Kehew, A.E.	176
Keller, J.B.	188
Kiersch, G.A.	141
Klasner, J.S.	41
Kloosterman, J.B.	1, 55, 76, 154
Kojan, E.	111
Koren, K.	55
Kranz, P.N.	236
Kranzer, H.C.	188
Krinov, E.L.	30
Krishnan, M.S.	87
Kumar, N.	146
LaMoreaux, P.E.	135
Lambert, A.	226
Lanphere	77
Ledbetter, M.T.	81
Lemoine, M.	122
Lidz, B.	155
Lindsay, J.F.	116

Author	Reference No.
Lipman, P.W.	136
Lirer, L.	84
Lonsdale, P.	149
Lonsdale, P.F.	105, 231
Lucchitta, B.K.	118
Lyell, C.	8
Lyell, K.M.	9
Mackin, J.H.	85
Malde, H.E.	175
Malfait, B.	149
Mamay, S.H.	170
Marinatos, S.	62
Masaytis, V.L.	39
McAfee, J.R.	54
McBirney, A.R.	200
McCrea, W.H.	53
McIver, R.D.	197
McLaren, D.J.	224, 225
McWhirter, N.	244
Meier, M.F.	208
Mercier, J.C.	96
Michel, H.V.	44
Middleton, G.	233
Mikhaylou, M.V.	39
Miller, D.J.	151
Moore, D.G.	125
Moore, J.G.	126
Moore, R.C.	167
Morris, S.C.	172
Mountjoy, E.W.	115
Mulholland, J.D.	52
Munno, R.	84
Murray, R.	233
Mutti, E.	113
Naeser, C.W.	83
Napier, W.M.	35, 48, 49, 134

Author Index

Author	Reference No.
Nelson, C.H.	148
Niklas, K.J.	220
Ninkovich, D.	63, 73, 81
Noerdlinger, P.	210
Norman, J.	42
Norrman, J.O.	152
North, R.G.	131
Obruchev, V.A.	166
Oehler, J.H.	213
O'Keefe, J.A.	46
Olson, E.C.	25
Olson, W.S.	180
Owen, D.E.	115
Palmer, R.	189
Park, C.	210
Penney, W.	189
Perkins, R.D.	147
Pescatore, T.	84
Pierce, W.G.	103
Plafker, G.	128
Plummer, P.S.	227
Pollack, J.B.	210
Porfir'ev, V.B.	108
Post, A	208
Powell, J.E.	55
Price, N.	42
Prostka, H.J.	124
Rampino, M.R.	59
Reid, G.C.	54
Reitan, P.H.	97
Rengarten, N.V.	214
Reynolds, S.A.	218
Ritter, D.F.	240
Robertson, P.B.	36, 37
Robinson, J.W.	132
Robinson, P.T.	243

Author	Reference No.
Rode, K.P.	106
Rodgers, J.	165
Rosell, J.	113
Roth, A.A.	101, 156
Rudwick, M.J.S.	13
Rupke, N.A.	229
Russell, D.A.	4, 55, 223
Sanders, J.E.	21, 146
Santacroce, R.	84
Sawatsky, H.B.	43
Scheidegger, A.E.	246
Schermerhorn, L.J.G.	117
Schindewolf, O.H.	221
Schultz, L.G.	230
Schultz, P.H.	133
Schulz, K.J.	41
Schumm, S.A.	239
Seguret, M.	113
Self, S.	59
Selivanovskaya, T.V.	39
Sequret, M.	113
Shea, J.H.	15, 20
Shinn, E.A.	137, 216
Shoemaker, E.M.	33
Sigleo, A.C.	212
Sigurdsson, H.	63
Simoneit, B.R.T.	231
Smith, B.A.	70
Smith, D.H.	218
Smith, H.T.V.	177
Smith, P.J.	45, 241
Smith, R.B.	80
Sonett, C.P.	186
Soter, S.	194, 195
Sozansky, V.I.	107
Spackman, W.	217

Author Index

Author	Reference No.
Spaeth, M.G.	144
Sparks, R.S.J.	63, 64, 71, 81
Spencer, E.W.	22
Spera, F.J.	98
Stanley, D.J.	157
Stanyukovich, K.P.	31
Stephenson, F.R.	53
Stewart, C.	143
Stipp, J.J.	155
Stockman, K.W.	137
Stokes, W.L.	162
Strelitz, R.	187
Strommel, E.	199
Strommel, H.	199
Swanson, D.A.	89, 92
Swanson, M.F.	155
Sweeney, R.E.	105
Symons, G.J.	58
Szekely, J.	97
Talent, J.A.	115
Talman, F.T.	211
Tazieff, H.	121
Thorarinsson, S.	68
Threet, R.L.	132
Toon, O.B.	210
Turco, R.P.	210
Twain, Mark	19
Twidale, C.R.	237
Vierbuchen, R.C.	139
Visher, G.S.	159
Vogt, P.R.	198
Voight, B.	119, 123
Walker, G.P.L.	66, 201
Walker, G.W.	92
Walker, R.	218
Warren, W.M.	135

Author	Reference No.
Warth, M.	55
Waters, A.C.	72
Watkins, N.D.	63
Watson, W.J.	60
Wedekind, J.A.	186
Weir, J.	114
Wetherill, G.W.	32
Wexler, H.	193
Wheeler, H.E.	90, 91
Whewell, W.	10
Whipple, F.J.W.	190
Whitten, R.C.	210
Wickramasinghe, C.	205
Williams, H.	82, 200
Williams, J.G.	33
Wilson, A.T.	209
Wilson, C.J.N.	201
Wilson, I.	71
Wolfe, R.F.	33
Wright, T.L.	89
Wyllie, P.J.	94
Yochelson, E.L.	170
Yokoyama, I.	61, 145, 191
Zeylik, B.S.	40

SUBJECT INDEX

Subject   Reference No.

Actualism 27,163
Agate Quarry, Nebraska 166
Air waves 30,48,49,141,189-193
Air waves, energies of 189,190
Alaska, muck deposits 207
Alaskan Earthquake (1964) ...119,128,
                              .130,144,192
Albedo 202,205,209
Anastomosing channels 174-179
Anhydrite 182
Annealing recrystallization of
  olivine 96
Antarctic Ice Sheet 209
Apollo object 32,33,186
Artifacts lithified in ocean
  sediment 216
Ash-flow tuff (see Pyroclastic flow
  deposits)
Asteroid (see also Meteorite) ..32-36,
                         44,185-189,224
Asteroid impact, effects of ...38,39,
   44,45,48-50,52,106,134,185-188,224,
                                    225
Asteroid impact, energy of 35-39,
                      42,44,134,185,186
Asteroid impact, probability of ...32,
                                 33,35
Asteroid impact, size of craters ..36,
                                    37
Atmospheric catastrophes ...44,48,51,
           198,200,202-207,209,210
Atmospheric cooling 44,49,54,73,
           199,200,202,204-206,210

Subject	Reference No.
Atmospheric dust	44,48,49,199, 201,205,206,211
Atmospheric heating	51
Atmospheric transparency	210
Avalanche (historic)	57,246
Axis change of earth	50
Basalt	87-93,104,198,243
Base surge deposits	72
Bassein submarine slide, Indian Ocean	125
Beach formation	138,152
Bering Sea, Alaska	148
Bermuda Triangle	197
Bias	1,3,4,14,20-24,27
Bioherms	115
Biotite, rapid crystallization of	219
Bishop Tuff, western United States	77,83
Blowouts from gas hydrate	197
Blue-green algae, artificial fossilization of	213
Bone breccia	166
Boulders (see Erratic boulders and Gravity flow deposits)	
Bouma sequence	148,156
Brain-washing	3
Breaching of natural dam	111,174,182
Breccia (see also Megabreccia)	38, 39,103,112-114,136,152,164,166,180
Brines, hot juvenile origin	107,108
Brito-Arctic flood basalts	198
Burgess Shale, British Columbia	172
Burrows	236
Calcite-compensation depth	51
Caldera	64,66,73-76,80,81,83,136

Subject Index

Subject	Reference No.

Caldera collapse 59,62
Caloris Basin (Mercury) 133
Cambrian 172,180
Cambrian-Precambrian boundary 180
Canyons (See Valleys)
Carbon dating 201
Carbonatites 69
Catalog of catastrophic
 processes 244-249
Catastrophe, word defined 6
Catastrophism, accepted 1-5,12,
            13,20,26,27,35,174,221
Catastrophism, rejected 2-4,8,9,
                       21-24,146
Catastrophist geologists 3,12-14
Cave Springs Cave, Virginia 139
Cave formations 232
Cementation of fossils (see
 Fossilization)
Cementation of rocks (see
 Lithification)
Channeled Scabland, eastern
 Washington 24-26,174,178,179
Channels (see Anastomosing channels)
Chert 104,164,213
Chile earthquake (1960) ..121,130,244
Chinese literature 201
Chlorophyll 220
Clams 236
Clastic dikes 101-103,114,127,
                           129,197
Coal 102,161,217,218,230
Coal balls 170
Coalification 217,218
Coconino Sandstone, Grand Canyon ..160
Coconut fossils 169

Subject	Reference No.

Collapse features 127,135,136,234
Colluvium 214
Colorado River flood (1905-1907) ..140
Columbia River (see Channeled
 Scabland, eastern Washington)
Comet impact 29,35,48,51,189,205,
 206,210
Compaction of sediments 102,127
Conglomerate 116,142,162
Conjecture 19
Cooling (see Atmospheric cooling)
Coquina 23
Coral-algal reef 233
Cordierite, rapid crystallization
 of 219
Cosmic ionizing radiation ...53-55,221
Cracks in unconsolidated sediments
 (see also Clastic dikes) 127,
 129,197,227
Crater Lake, Oregon 82
Craters, impact of cosmic body (see
 Asteroid impact)
Craters, volcanic (see Caldera)
Cretaceous-Tertiary boundary ...44,198
Crete 62-65,145
Cross-bedding 159,160
Crystallization of minerals,
 rapid 96,219
Currents (see Ocean currents)
Cyanide poisoning 51
Cypress wood, artificial coalifica-
 tion of 217
Dam, breaching of 111,174,182
Dating (see Radiometric dating)
Deccan Traps flood basalt (Cretaceous,
 India) 87,88,198

Subject Index

Subject	Reference No.

Deep ocean drilling 243
Delta formation 19,140
Deposition (see Sedimentation)
Diamond 94,95
Diapiric processes (see Intrusive processes)
Diatomaceous earth 171
Diatreme (see Kimberlite)
Dikes (see also Clastic dikes, Kimberlite) 97
Dinosaurs 4,44,223
Dogma 1,8,12,20
Douglas Creek, Colorado 239
Dry fogs from volcanic eruption ..200
Duke Island ultramafic complex, Alaska 99
Dunes 149,159,160,238
Dust (see Atmospheric dust)
Earth gas 194
Earth's mantle 94,108
Earthquakes 29,30,48,102,113,114, 121,124,125,128,130-134,144,150, 151,244-246
Edgewise conglomerate 142
Emotionalism 5,22,24
Environment of deposition (see Sedimentation)
Erosion, canyons 132,140,153,154, 174-176,239
Erosion, cave 139
Erosion, coastal 152
Erosion, tsunami (see Tsunami, erosion)
Erratic boulders 114,181,206
Eruption of gases 127
Erosion, flood (see Flood, erosion)

Subject	Reference No.
Evaporites (see also Salt)	182
Evolution, organic	222
Evolutionism	17
Explosion of gases	127,194,195
Explosions	28-31,48,56-58,127, 188-191,193-195,244
Extinction of life	44,48,49,51,53, 204,205,221-225
Eyewitness reports	30,52,60,127,143
Fault breccia	103
Faulting (see also Overthrust faulting)	108,114,127,129,130, 132-134
Faunal discontinuities	4,44,48,49, 51,221-224
Ferromanganese deposits	241
Fire and flames	30,194
Fish, mass mortality of	247
Fish Creek drainage, San Diego County, California	132
Fission reactor	242
Fissures in unconsolidated sediments (see also Clastic dikes)	127,129, 197,227
Flavonoids	220
Flocculation of clay	230
Flood, erosion	24,140,174-180
Flood, sedimentation	24,64,140, 153,156,158,161,162,174-176,180,181
Floods (historic)	57,60,111,127, 137,138,140,141,144,145,147,151, 153-155,240,245,246
Floods (prehistoric)	24,25,48,49, 157,158,161,162,167,170,174-182,224
Fluid pressure in catastrophic sliding	120

Subject Index

Subject	Reference No.

Fluorite, rapid crystallization
  of 219
Flute casts 142
Fogs from volcanic eruptions 200
Foraminifera 155
Fossil cemeteries 166,207
Fossil forests 167,228
Fossil reef 233,234
Fossilization ...165-168,170-172,212,
              213-218,220,228,229
Fracture formation in rock 95,97
Frame-building organisms 233,234
Franciscan Formation, California .104
Freezing rain 205,206
Fusulinids 173
Gas eruptions 127,194-197
Gas hydrates 197
Geology, history of 3,4,7,10,
            12-14,17,20,21,24
Gibraltar, Straits of 182
Giordano Bruno, lunar crater 52
Glacial deposits 116,117,180
Glaciation ...49,155,174-177,181,202,
              205,206,209
Glaciers, rates of movement 208
Global cooling (see Atmospheric
  cooling)
Gorges (see Valleys)
Graded bedding (see Turbidites)
Gradualism 21
Grand Canyon, Arizona 112,160
Gravity anomaly 41
Gravity flow deposits (historic) ..57,
           64,109-111,148,150

Subject	Reference No.
Gravity flow deposits (prehistoric)	112-118,125,156, 157,228
Gravity slide deposits (historic)	119-121
Gravity slide deposits (prehistoric)	103,121-126
Great Reform movement of 1832	14
Guaymas Basin, Gulf of California	105,231
Guiana Shield, South America	76
Gulf of Mexico, seiching of	19,144, 155
Hartford Dike slide	120
Hawaiian Ridge, submarine slides	126
Heart Mountain rockslide, Wyoming	103,123,124
Heat wave	30
Hecho Group, Pyrenees, Spain	113
Herring fossils	171
Huascaran rockslide, Peru (1970)	110
Huckleberry Ridge Tuff, northern Rocky Mountains	80
Hurricane Carla (1961)	138
Hurricane Donna (1960)	137,147
Hurricanes (see Storms)	
Hydrate of natural gas	197
Hydrothermal crystallization	104-108,213
Hydrothermal deposits	104-108,231,241
Ice age	49,174-177,181,202,203, 205-207,209,
Ignimbrite (see Pyroclastic flow deposits)	
Imbrium Basin (the Moon)	133
Impact breccia	38,39

Subject Index

Subject	Reference No.

Impact craters (see Asteroid impact)
Impact frequency 32-35
Impact structures ..33,36-38,41-43,49
Intrusive processes (hot and
  cold) 94-103,114,127,129,197
Io (Moon of Jupiter) 70
Iridium, extraterrestrial source
  of 44
Isigaki Island tsunami (1971) 244
Ishim impact structure
  (Kazakhstan) 40
Island Park caldera, Idaho 80
Kanamori scale of earthquake
  intensity 130,244
Kimberlite 69,94-96
Krakatau eruption, Indonesia
  (1883) 56,58-60,62,73,191,193
Lake Taupo caldera, New Zealand
  (186 A.D.) 66,201,244
Lake Toba caldera, Sumatra 73-75
Lake, drainage 111,174
Lake, formation 111,127,140
Lake, seiche 143,144
Laki lava flow, Iceland (1783) ...56,
                      67,68,89,200
Laminated sedimentary rocks ..114,226
Landforms ..24,152,153,174-179,237-239
Landscapes, evolutionary theory for
  origin 237-239
Landslide (see Gravity slide deposits)
Landslide (historic) 57,119-121,
                      141,150,246
Lava Creek Tuff, northern Rocky
  Mountains 80
Lava flows (historic) 67-69,244
Lava flows (prehistoric) ...87-93,198

Subject	Reference No.
Lava flows (submarine)	93
Leaf fossils	220
Limestone	113,115,135,139,164,173, 233,234
Linear magnetic anomalies	243
Lithification	101,163,174,215,216
Lituya Bay rockslide, Alaska (1958)	151
Loch Tays seiche, Scotland (1784)	143
Long Valley caldera, California	77,83
Lunar volcanism, origin of tektites	46
Magma ascent velocities	95-98
Magnetic field reversal	48,49,243
Mammals	4,207
Mammoths	4,206,207,214
Manganese nodules	241
Mantle of earth	94,108
Marine fossils within terrestrial deposit	142,170
Mars, landslide deposits	118
Mass mortality of marine organisms	247
Mayunmarca rockslide, Peru (1974)	111
Meandering streams	184
Mediterranean Sea	162
Megabreccia (see also Breccia)	112-115,136
Mercury	133
Mesa basalt, northwestern United States	90-92
Mesa Falls Tuff, Idaho and Wyoming	80
Metaphysical beliefs	12
Meteorite (see also Asteroid)	29,31

Subject Index

Subject	Reference No.

Mid-oceanic ridges 243
Minoan civilization 62-65
Mississippi River 19,127,155,178
Mono Lake, California 235
Moon 46,52,133,180
Mount Mazama, Oregon 73,82
Mount St. Helen's eruption, Washington
 (1980) 57
Muck deposits (frozen soils) 207
Mud flows (see Gravity flow deposits)
Mudcracks 227
Multiple-working hypotheses 2
Myths and legends 35,155
Nappe formation 121,122
Navajo Sandstone, Utah 159
Neocatastrophism 5,221
New Madrid earthquake, Missouri
 (1811,1812) 127,195
Nickel deposit 38
Night sky glow 211
Nitric oxides 49,53,54,210
Nuclear explosion, air wave generated
 by 193
Nuclear reactor 242
Ocean currents 138,142,149,159,
 160,169,173
Ohio River 178
Oil 43,231
Oklo uranium deposit, Gabon Republic,
 West Africa 242
Olistostrome 113,125,126
Olivine, recrystallization of strained
 crystals 96
Organic binding of sedimentary
 rocks 233,234
Orientale Basin (the Moon) 133

Subject	Reference No.

Overthrust faulting, mechanism of 103,119-124
Owens Valley earthquake, California (1872) 194
Oxygen isotopes 105,155
Ozone 48,49,53,54,210
Paleontology 5,221,222
Paleosurfaces 162,237
Paraconglomerates 142
Pediment 132,162,238
Penecontemporaneous deformation ..101-104,142,164,173
Peridotite 94-96
Permafrost 214
Petrified Forest National Park, Arizona 167
Petrified wood 167,168,212,228
Petroleum 43,231
Phosphorites 163
Photosynthesis 54,205
Pipe (see Kimberlite)
Plains 237
Plankton 51
Plate tectonic theory 48,243
Plato's date of the Flood 155
Poisoning 51,56,198
Politics 14
Pollution 198
Polonium halos 218,219
Polystrate fossil 168
Popigay impact structure, Siberia ..39
Potassium-Argon dating (see Radiometric dating)
Pseudofossils 213,234
Pumice 59,64,66,83,145

Subject Index

Subject	Reference No.

Pyroclastic eruptions
  (historic) 56,57,59,60,63-
            66,70,74,75,82,244
Pyroclastic flow deposits
  (historic) 57,59,71
Pyroclastic flow deposits
  (prehistoric) ..56,66,72,77-81,83-86
Pyroclastics, theory of eruption and
  flow 57,59,66,71,72,78,81,86
Radioactive fossil bones 55
Radioactivity 53
Radiohalos 218,219
Radiometric dating 92,201,218,241
Rapid burial 103,156,165-172,179,
                            228,229,236
Rate of displacement on faults 49,
      103,113,119-124,127,128,132-134
Rates of lava eruption 68,89
Rates of sedimentation (see Rapid
  burial)
Reefs 115,233,234
Reelfoot Lake, Tennessee 127
Relict landforms 178,237,238
Remnant magnetism of seafloor
  basalt 243
Rhinoceros fossils 166
Richter scale, modification of ...130,
                                  244
Ridiculing, scoffing 1,2,221
Ring volcanoes 76
Rip-up clasts 164
Rockslide 103,110,111,122-124,
                              151,246
Roman literature 201
Root horizons 229

Subject	Reference No.

Rosa basalt flow, Washington
  (Miocene) 89
Rotation of earth 49,50
Sahara, Africa 178,238
Salt, hydrothermal origin of ..106-108
Salt domes 100
Salt intrusion 100
Salton Sea, California, formation
  of 140
Sand dunes (see Dunes)
Sandstone 146,148,149,156-160
Santorini eruption (see Thera)
Scablands 24,174
Sea cliffs 152,238
Sea level change 50,206,238
Seafloor spreading 243
Seamounts 126
Sedimentation, environment of .56,57,
    59,60,64,70-72,78,103-107,113-122,
    127,137-140,142,145-150,152,153,
          156-173,182,214,228,229
Sedimentology, dominated by uniformi-
  tarianism 23,27,146
Sedimentology, neocatastrophism
  gaining ground in 5,27
Sedimentary structures in igneous
  rocks 99
Seiche 127,142-144
Seismic waves (see Earthquakes)
Severe weather 199,205,206,245,
                              246,249
Shale 104,114,161,172
Shale partings in coal bed 161
Shatter cones 38,39
Shell breccia (see Coquina)

Subject Index

Subject	Reference No.

Shimabara Bay, Japan, rockslide-
 generated sea wave (1792) 151
Shinarump Conglomerate, Colorado
 Plateau 162
Siberia 28-31,39,210,214,
Silica gel 104
Silica mineralization 104,167,212
Sinkhole 135
Sky glow at night 211
Snake River Plain, Idaho 175
Sodium carbonate lava flow 69
Soil 207,215,228,230
Solar flare 54
Souris spillway, Saskatchewan 176
South America, gigantic calderas ...76
Speculation 1,8,11-13,19
Spillway 174,176
Spokane Flood (prehistoric) 174
Stalactites 232
Stalagmites 232
Stigma associated with
 catastrophism 4,21,22,145
Stigmaria (fossil rootlike organ) .229
Storms, sedimentary products of ...23,
      137,138,146-148,158,164,170,173
Storms, surging waves ...137,138,148,
                                   170
Storms, used to interpret ancient
 sedimentary rocks 23,138,146,
                            161,164,173
Strained olivine crystals 96
Straits of Gibraltar 182
Stratosphere 54,210
Stream drainage, diversion of 132
Stromatatactis (sedimentary collapse
 structure) 234

Subject	Reference No.

Submarine canyons 142
Submarine fossilization and
  lithification 215
Sudbury impact structure (Ontario) 38
Supernatural, unnecessary in
  defining catastrophe 6
Supernatural (see Metaphysical
  beliefs)
Supernovae 53,55
Suppression of evidence 14
Surging glaciers 208,209,246
Surtsey eruption, Iceland
  (1963) 152,215
Suva earthquake and tsunami, Fiji
  Islands (1953) 150
Synaeresis cracks 227
Talc, hydrothermal origin of 105
Tambora eruption, Indonesia
  (1815) 56,61,73,199
Tapeats Sandstone, Grand Canyon ...112
Taupo eruption, New Zealand
  (186 A.D.) 66,73,201,244
Tectonic events ...96,103,104,119-122,
                            127-129,134
Tektites 45-47
Tempestite 23,164,169,173
Tephra deposits 63-65,74,75,82
Terraces 152,153,176,214,239,240
Thera (Santorini) eruption, Aegean
  Sea (1500 B.C.) 62-65,73,145
Tidal waves (see Tsunami)
Tides 180,181
Tillite 116,117
Toba eruption, Sumatra
  (Quaternary) 73-75,81

Subject Index

Subject	Reference No.

Transport mechanism of enormous rocks 109,112
Tree, upright burial of 168
Trilobites 172
Truckee River, Nevada 240
Tsunami, erosion 142,169
Tsunami, sedimentary products of 142, 145,158,169,170,244
Tsunamis (historic) 58-60,62,64, 141,142,144,145,150,151,244
Tsunamis (prehistoric) 48,49,114, 142,158,169,170,224
Tsunamis, theory of generation ...185-188,196,246
Tufa, rapid deposition of 235
Tuff (see Tephra and Pyroclastic flow deposits)
Tunguska explosion, Siberia (1908) ...28-31,189,190,193,210,211
Turbidites 138,142,148,156-158, 169,226
Turbidity currents 142,150,156, 169,197,226
Turnagain Heights landslide, Anchorage, Alaska (1964) 119,123
Ultrabasic rock, layered 99
Ultraviolet light 48,49,53,54,210
Unconformity 162,179,180
Underclay below coal beds 230
Underfit streams ..174-176,183,184,238
Uniformitarian geologists 3,4, 12-14,22,24
Uniformitarianism, accepted 9
Uniformitarianism, rejected 1-3, 5,7,10-20,23,26
Uranium halos 218

Subject	Reference No.

Uranium ore deposit 242
Vaiont slide, Italy (1963) ...122,141
Valleys 132,140,142,153,154,
174-176,182-184,238
Varves 226
Vine-Matthews hypothesis 243
Volcanic breccia 103
Volcanic dust 199,201-204,207
Volcanic eruptions, energy of ...57,61
Volcanic gases 124,198,200
Volcanism 48,46,56-93,198-204,
223,244
Water waves 48,137,138,141-145,
148,150,151,186,188,244
Weather, severe events 137,138,
147,199,200,205-207,210,245,246
Welded tuff (see Pyroclastic flow
   deposits)
Westphalia Limestone (Oklahoma, Kansas
   & Missouri) 173
Williston basin (Saskatchewan,
   Manitoba & North Dakota) 43
Wind 48,206
Wood, artificial fossilization
   of 212,217
Wright Dry Valley, Antarctica ...177,
178
Xenoliths 96,98,215
Yellowstone caldera, Wyoming 80
Yellowstone National Park,
   Wyoming 80,212,228
Zion National Park, Utah 159